Rivals in Restraint
Kathleen Haley

Cycling Romance Press—Berkeley, CA
ISBN: 979-8-9861474-0-6
Library of Congress Control Number: 2022908114
Title: *Rivals in Restraint*
Author: Kathleen Haley
Digital distribution | 2022
Paperback | 2022

Dedication

To Maurice,

Whose sound judgement kept the characters rooted in this world.

Acknowledgments

A huge thanks to all those who read this manuscript and gave such immensely helpful critiques: Maurice Morton, David Haley, Anna Haley, Caleb Haley-Moskalev, Annie Goldman, Raja Sengupta, Marta Gonzalez, Pablo Guevara, Tanya Pollard, Alex Adams-Leytes, Laura Jane Wey, and Peter Schwartz.

"Thou and I are too wise to woo peaceably."

— Benedick, *Much Ado About Nothing*

Chapter One

an was finding Jay to be the second sort of date. She had had two kinds of dates over the last month: the men who held forth volubly about their jobs, hobbies, travels, and family, without showing a modicum of interest in her own; and the men who preferred letting awkward silences build to helping the conversation along, and who forced Nan to do the heavy lifting. Unfortunately, the cursory interest this second sort of man showed in her insights rarely led to follow-up questions or spin-off material that Nan could use as she cast about for ways to make it look as though they were having an easy, enjoyable exchange. Men like Jay made Nan feel as if she had returned to college teaching mode and was trying to keep the discussion moving along at all costs.

"So how was your company affected by the pandemic?" she asked Jay, who had co-founded a cyber security company in San Francisco.

"Like a lot of companies, we ended up going completely remote and will likely keep our employees at least partly remote from now on," said Jay. "It's a small company though—only fourteen people. As far as our business goes, it hasn't suffered from the pandemic, but neither has it grown that much."

As she tried the cold, rather bland, but wildly expensive trout ceviche she'd ordered, Nan counted the slow seconds in her head before Jay bounced the ball back in her court.

Finally, he said, "What about you?" She suspected he'd forgotten what she did for a living.

"It's been super challenging to be a musician for the last year and a half. For a while, I gave Zoom classes to those of my students who already have a strong basis in piano, but ultimately piano is not a great instrument—actually, no instrument is great—for learning over a screen in separate locations. Then there are all the complications involved in trying to play together as a group at a distance or to record albums. Things have eased up since the vaccine was rolled out, but all of us musicians feel we've lost a lot of time—like the rest of the world."

Jay nodded.

"How's the mixed fish?" asked Nan.

"It's okay. You want to try some?"

Nan shook her head. "No, thanks."

"Why are you sitting like that?" said Jay.

"How do you mean?"

Jay smiled for the first time. "Like you're at attention or attached to a board."

"Oh. My mom always made us sit up straight at the table," said Nan. "If we slouched, she'd emphatically straighten her own posture to remind us."

"Hmm. Interesting."

Dinner wore on, and Nan began to feel tired, although it was barely 9. She reflected that at least the first kind of date taught her things about the world; the second kind made her only conscious of how much time she was losing. Jay insisted on tabulating his own items on the bill and splitting their cards unevenly, despite the fact that the second glass of wine Nan had ordered—largely to combat boredom— she'd ended up pouring mostly into Jay's own glass, and their ceviches had only differed in price by a few dollars. As if sensing that it was now or never to prove himself, Jay began to wax enthusiastic about vampire movies and Netflix shows he thought they could watch together.

As he drove Nan the mile back to her apartment down College Avenue, he turned several times to her with an energy he hadn't revealed in the restaurant, and said, "I really enjoyed tonight. Maybe next time you can come over to my place and we can order in and watch Netflix."

Nan nodded, amenably kissing Jay's cheek as he leaned in to say goodnight. She could never screw up the courage to be direct at the end of a first date. She always had to call the man back another day and ask if they could be friends (by which she really meant she hoped the man wouldn't take offense at her not wanting to see him again). Nan admired women who could simply turn to their date and say, "I'm not interested in seeing you again. I wish you the best in finding what you're looking for."

"How was it?" Nan read Margot's text as soon as she got in the door.

"You up for a chat?" Nan texted back.

The phone rang immediately. "Hey, sweetie." Margot's voice brimmed with the spirit Nan had missed for the last two hours. "Did you just get back?"

"Yeah." Nan felt defeated.

"Oh. Okay, which kind of date was he?"

"The second."

Margot sighed. "Did he at least look like his picture this time?"

"I guess so. But it didn't matter, because all the rest was so off. Marg, it feels like such a huge waste of time. How do you keep at it?"

"You know what I've always said: dating is the one train no one who's single can afford to get off of preemptively."

"And what's the last stop on that train?" asked Nan, hopefully.

"Celibacy, probably," Margot returned somberly.

The next afternoon, as she rode her bike back from the repair shop, where she'd had a new phone battery installed, Nan revolved the pros and cons of internet dating. In theory, it suited the lives of busy people who wanted to be more in control of how they met their romantic partners and the kinds of relationships they had with others. In practice, Nan had found that the same amount of pure luck applied with internet dating that applied in real-life romantic encounters. Maybe online you needed even more luck on your side, since fewer people were still online than out there in the world waiting to be met. As far as the matching algorithms went for the various online dating sites, Nan felt that ultimately there was so much intangible, immeasurable humanness to each person that the virtual world couldn't capture, that relatively speaking it was still a crapshoot. If only she could enjoy the crapshoot more.

As she passed the courthouse, she encountered hundreds of demonstrators, all of whom rode bikes, held up signs, and shouted rallying cries about police, cycling rights, and racial justice. They were mostly in their twenties and thirties, and were riding everything from fold-up bikes and commuters to mountain bikes and glammed-up cargo bikes. Nan instinctively sheared off from the protesters and averted her eyes, guiding her own bike around the western side of Lake Merritt, leaving the crowds behind, and hastening towards home. It wasn't until she reached Broadway, when it was decidedly too late to do anything about it, that she regretted not stopping and asking a participant what the protest concerned. She was curious, and more than a little guilty about putting blinders on, but her usual fear of looking at things directly, of losing control, of missing out on time allocated for other activities, won out, as always, and she bent her head down, determined to get back and practice piano.

When she came in the door, her sister Bess called.

"Bertie, Pooh-bah, and I are headed to Tennessee Beach for a light hike. You want to come?"

Nan thought quickly. She briefly envisioned a windy late afternoon on the beach watching Bess's dog Pooh-bah tear wildly through the sand as Bess's six-year-old son Bertie chased him gleefully, the tide rose steadily, and the seagulls screeched and dipped. It would be a rare Wednesday escape, a proposition even more tempting than the urge to slow down and take in the demonstrations. But she couldn't justify skipping her three hours of practice.

"I'd better stay put. But I'll see you guys on Friday afternoon."

Nan had become expert at denying herself joys like these, just as she was deft at numbing her senses to the outside world and applying her tunnel vision to every discipline she engaged in. She had become adept at compartmentalizing time for her various social, professional, creative, and athletic activities. And she was acutely aware of time at every moment—from the number of seconds it took to fill the kettle with water for one large cup of tea and the number of steps it took to reach the vegetable market on the corner, to the number of breaths she took when sleeplessly tossing and turning at night and the number of seconds she saved climbing the hill on her bike on any given day. She was a pastmaster at counting and accounting for time. Yet somehow, the better she got at managing small increments of time, and the more she kept a vigilant eye on the short term, the more depressed she was, and the less she accomplished in the long term.

Nan longed more than anything else to play Schubert's gloriously complex sonata in B-flat major this afternoon. Instead, she played everything but Schubert. She warmed up with an hour of the least pleasant exercises she could think of in Hanon—the ones that focused on trills and arpeggios—and then worked her way through Bach's sobering keyboard sonata in D-minor and a restrained Haydn sonata. By the time her three hours were up, she could only eye the book of Schubert sonatas with self-reproach and remorse.

After fixing a salad and microwaving a potato for dinner, she copied out a poem by John Donne titled "A Nocturnal upon St. Lucy's Day," in the aim of memorizing it by the actual St. Lucy's Day in December. Then she tried to sketch Pooh-bah from photos she had of him on her phone, planning to compose a watercolor of a family of dachshund-Chihuahua mixes who looked just like him dancing around a fire. As usual, she lacked inspiration, and the hour she spent drawing felt forced, even as her obsession over small details—like Pooh-bah's whiskers or the curve of his toenails—kept her from capturing the spirit of his movements and his overall form. Frustrated, she set the drawings aside.

From the corner of her desk her old staffed notepad and a folder of notes glowered at her accusingly. Would tonight be the night she finally opened them up? She had considered them at various times of the day for the last six months. But it was too late now to begin anything that big—it was already almost 9. To resume work on her quintet she would need to start early in the day, as soon as she awoke. Why wasn't she working on her composition every morning from 8 to 11? She thought of how she'd spent Monday and Tuesday morning— looking up resistance exercises for the boot camp she taught, reading articles about fitness, and watching workout videos. Monday midday she'd spent another two hours doing her own weight workout based on the ideas she'd generated, and Tuesday midday she'd run up the fire trail. This morning and midday, of course, she'd ridden the bike with Thomas. Maybe tomorrow she would work on the quintet at last.

She spent an hour and a half checking email and renewing her piano teaching posts on Nextdoor and Craigslist, and went to bed, turning out the light at 10:30. Waking the next morning at 7:30, she remembered she had her first therapy appointment at 10 with a man named Naseer Zadeh. There was no point in trying to get any serious composing done in just one and a half hours. She would devote Friday morning to it.

Naseer was tall, slender, dark-haired, and soft-spoken, with an affable demeanor. After a few polite preliminaries, he said, "What brings you here?"

"I guess a few things," said Nan, leaning forward and focusing on the bowl of smooth grey pebbles on Naseer's table. "I want to rediscover the comic side of life. And I want to stop letting my perfectionism get in the way of accomplishing something decent. And I want to stop obsessing over time so I can regain the ability to appreciate the present moment."

Naseer nodded. "Is there a reason you're coming here now?"

Nan expelled a breath she hadn't realized she was holding. "My mom died in May 2019 of metastasized breast cancer. She was only 64. I don't think anyone—my sister, my brother, or my dad—has really been able to laugh or joke around since then. I was especially close to her—we talked every day on the phone for at least half an hour—and instead of getting easier over the last two and a half years, it's gotten harder. I lost my job as a music lecturer at Drummond University in

New York in August 2019. Then I moved out here, the pandemic hit, and it was extra tough to be a musician and music teacher. And recently I've been feeling guiltier and lonelier than ever."

"Why do you feel guilty?"

"My mom gave me generous financial support ever since I was in college. I wouldn't have been able to keep up all my extracurricular pursuits so avidly all this time without her help. I'm worried that in missing my mom, I'm largely missing her financial backing, and that my relationship with her was mostly selfish. I also feel guilty about how much time and money I've squandered over the years."

"In what way?"

"Drummond ended my lectureship and hired someone to a tenure-track position in my place because I hadn't recorded, composed, or published enough articles since joining the faculty. I spent all my time teaching, doing athletics, writing poetry, singing, and painting. I taught undergraduates music appreciation, history, composition, and theory, and I focused on learning good teaching techniques and rounding out my teaching portfolio. I genuinely loved teaching college and I'm curious about everything, so it wasn't hard for me to throw myself into all of these subjects. But I willfully deferred putting together the pieces of a quintet I've been composing over the years—if I'd completed it, I might've kept my job. I'm a horrible procrastinator. Most of my graduate school peers have gone on to build successful careers, and I think it's because they all realized early on the value of time and how to make the most of it. I'd give anything to teach college again. I've been applying every day to college positions since moving out here."

"And why do you feel lonely?"

"I miss my mom, and I miss the lightness in life that her presence brought. Since she died, my perfectionism has gotten much worse. And I think that my perfectionism is keeping me from finding a romantic partner."

Naseer jotted something down in his notepad. "How so?"

Nan looked at him directly. "I have a hard time accepting imperfection in myself and in people I'm close to. I haven't dated anyone in three years now."

"Is that because you've met people but haven't found them date-worthy?"

Nan nodded, ashamed. "I'm afraid so. Two months ago I started online dating, but the nine men I've met in person didn't measure up. I suspect I'm being overly judgemental. But I can't help it."

Naseer looked at his pad and returned to a previous thread. "You said you wanted to 'regain the ability to appreciate the present moment.' Would you say that's also connected to your mom's death?"

"Definitely. I used to take joy in details like the dew on the ferns when I rode the bike, or the rustle of chestnut leaves when I walked around the neighborhood, or the phases of the moon when I went out at night. My ability to notice things like that has steadily diminished since my mom died. Nowadays, instead of appreciating details, all I can do is obsess over them—which doesn't feel like the same thing at all. Like when I sketch, details act as stumbling blocks—no longer as sources of inspiration. And during my last year of teaching at Drummond I became too picky in my grading. I couldn't get over the small details and look at the larger picture. All I saw was students' mistakes and shortcomings. And now, as a piano teacher, it's an ongoing battle for me to lower the bar in my own head so that my students can meet me halfway."

"This week would you like to focus on enjoying details again?" asked Naseer.

"That sounds great."

"How about if, three times a day, you take one minute to try to use as many of your five senses as you can to notice—really *notice*—the things around you. If you're taking a walk, stop and breathe in the smells. Turn around and look at everything within ten feet of you from different angles. Listen to the sounds both near you and further away. Touch something to make the moment full of meaning."

"I'll try." Nan was hopeful.

Naseer smiled beneath his mask. "Remember, it's only too human to fear accepting any new self that's too big a change from your old self. All we're doing is tricking your mind into accepting a new self by slow degrees."

♪ ♪ ♪ ♪

As Nan rode home on her e-bike, she thought about the musical, artistic, and athletic skills that for so long she'd been able to sustain in a tenuous state of equilibrium, yet which had cost her, on many occasions, her sanity—in the form of manic episodes—and, at the critical juncture of promotion, her university teaching career. Just as she feared relaxing into any moment, she was afraid of devoting herself wholly to the composition. She had focused single-mindedly on one project before—when writing her dissertation or an article for a music education journal—but now she worried constantly that if she relaxed her attention to piano practice, cycling, weightlifting, poetry, singing, or drawing, she'd lose all the expertise she'd worked so hard to build up over the last twenty-some years in all those areas.

She often saw herself as a soldier marching in lockstep to her own self-imposed pursuit of excellence. Her motivation for following this relentless onward march derived equally from habit and from the desire to separate herself off from the common herd. In the realm of exercise, her habit—or, rather, addiction—had developed from her perennial anxieties concerning her body. She and her mom, Gwen, had had many conversations about her pursuit of excellence, which had begun during her undergraduate years at Brandeis. Nan tended to take everything good to an extreme, including weightlifting, running, and cycling. At one point, when she was spending five hours a day in the gym and, on her off days, riding three hours or more on the bike or running twelve miles, Gwen had asked her, "What's your exit strategy?" Nan had frankly acknowledged that something had to give, if she was to finish her dissertation and get a teaching job, all while maintaining her piano skills.

But instead of letting up on extracurriculars and athletics, she had simply opted to "chase daylight," as she called it. From age 20 to age 34, she'd fallen into a pattern of rotating around the clock in her sleep, as if she was constantly traveling west one time zone a day for twenty-four days. On day one, she might go to bed at 10 p.m. and get up at 7 a.m. But by day twelve, she'd be going to bed at 10 a.m. and getting up at 7 p.m. This would continue, until, on day twenty-four, she'd return to the same sleep schedule as day one. In chasing daylight, she managed to eke out an extra hour every day, but it became a habit she couldn't break, and she discovered it was even a disorder in its own right, called non-twenty-four-hour sleep-wake disorder. According to a psychiatrist she'd seen at Harvard while getting her doctorate, it was common for people like Nan with bipolar-two disorder also to have a sleep disorder.

A year ago, substitute teaching had finally forced Nan out of this rut. This was before she'd yet built up her current base of piano students, when she'd supplemented her income subbing at high schools in Oakland and Berkeley across various subjects. The job required her to get up at 7:15 every morning and, with the help of Ambien, which she took when she woke up four hours into every night and couldn't get back to sleep, she at last established a regular diurnal schedule after three months of intense struggles and several near-breakdowns. From the standpoint of all who knew her, this was nothing short of a miracle. Nan imagined how proud her mom would have been to know that she had finally achieved normality in sleep.

But like an addict, she had to be hyper-vigilant about her newfound sleep "sobriety," as she conceived of it. Several nights of staying up past 10:30 p.m. could throw everything out of whack for weeks. Similarly, she feared what might happen if she drifted far beyond 7:30 a.m. for a wake-up time. She had long kept a journal in which she religiously recorded every time she went to bed and woke up. Now it was sacrilege to diverge too much from her successful nine-hour window.

Nan had acquired a new student that week named Melanie, who, at ten, was younger than most of Nan's students, but who was emotionally mature and a very gifted pianist. Nan never felt overjoyed teaching piano the way she'd joyfully embraced college teaching; one-on-one coaching never came naturally to her the way lecturing and facilitating class discussions did. But she knew she was a good teacher, largely because she channeled the techniques of her own beloved former piano teacher of twelve years, Lilian. As she coached her sixty-six-year-old student Linda through a fugue from Book One of Bach's *Well-Tempered Clavier*, Nan smiled to think how thoroughly she remained a student of Lilian, even after all these years. Lilian had always taught her students to aim for three things that made music study universally interesting: the ability to play with other people; the ability to sightread; and the ability to enjoy playing for one's entire life. Not only had these tenets shaped how Nan taught undergraduates, but they had remained central to her pedagogy as she gave one-on-one lessons.

Nan rode her e-bike over to the gym, picked up the bag of supplies for her boot camp, and rolled to the nearby elementary schoolyard where she set out an agility ladder, a row of exercise mats, two neat piles of jump ropes and resistance bands, and the speaker she used to project her playlist of classic rock, eighties rock, and Latin music. Six fit twenty-somethings showed up that evening, avidly following Nan through a series of explosive bodyweight exercises, partner-assisted activities, weighted movements, agility drills, and mat exercises.

"And five—four—three—two—one—okay! That's it," Nan counted down, to the relief of everyone doing the plank progressions she had devised.

"Real-time feedback: this kicked butt," said a newcomer named Kevin. "I'll totally be back again next week."

"Thanks," said Nan, cheeks flushed with exertion. "I look forward to seeing you again."

In all her career in fitness—that is, all ten years that she had devoted herself to serious resistance training—Nan had never had such consistently positive responses to her approaches to fitness and to her muscular body type as California had given her. Perhaps it was the conservatism of the Boston area or the superficiality of New York that had elicited the copious ambivalent responses she'd had from people in both of those places. A Russian photographer she'd dated a few times in Cambridge had once told her, "I want to take pictures of those legs. Not that anyone would ever believe they're not a man's legs." She had been walking to the gym one day when a group of young people had called out from a car, "Shit, look at that girl's legs. No . . . wait! It's a guy! Oh shit!" And she had indeed attracted no small number of gay or bisexual men through the gyms at which she worked out, some of whom had eventually become good friends. Then there were the staid older ladies of the Cambridge and Brooklyn streets who had given her ice-cold twice-over looks that could kill on impact. There were also the first dates in which the men had told Nan something to the extent of "your biceps alone could probably crack my legs, no problem." These latter comments were no doubt meant as compliments, but they turned Nan off immediately.

Nan biked the equipment back to the gym and then rode home, mentally replaying the other clients' responses over the course of the hour. One girl had said, "Burpees, huh?" in a doubtful voice that suggested one could push her only so far. Another guy had frankly appreciated the ladder work as drills he'd done during soccer practice in college. The sprinting towards the end of the workout had, by all accounts, felt like torture—but such a rewarding form of torture!

After dinner, Nan once again ventured a glance towards her dormant composition, a collection of bits and pieces of themes and phrases she'd begun six years ago and made halfhearted progress on ever since. Would tonight be the night she would finally open the folder and notepad? But she was determined to memorize the first two stanzas of

Donne's poem, and she wanted to map out the composition of her watercolor. The quintet could wait until tomorrow.

Friday morning, Nan's inbox overflowed with alerts for music teaching positions, and she spent until 1 applying to every remotely promising job, from part-time adjunctships to full-time lectureships. Then she went on a run up the UC Berkeley fire trail, mulling over the likelihood that she would land one of these positions. Music was, for her, only one of many passions; she was beginning to realize, for a successful academic career, it needed to be her sole passion. She knew she lacked neither energy, focus, nor self-discipline. It was merely her career ambitions that had tapered off since graduate school. She fully blamed herself for not advancing to an assistant professorship. Nor did she begrudge others their success. None the less, she often wondered how it was that so many people—in her field or other academic disciplines—could isolate one or two closely related areas of interest to pursue, present them efficiently and attractively to the world, and steadily progress in their profession as a result. Colleagues often played off their dedication to work by complaining of procrastination, laziness, the distractions of family, and lack of inspiration. But Nan had learned the hard way to read between the lines: each person's supposed sluggishness and lack of drive really hid a ruthless desire to succeed in a cut-throat field at any cost.

As she ran up the steepest portion of the route, a connector between two flatter parts of the fire trail, Nan remembered how after Drummond had let her go, she had determined to make a change. Consulting with her mentors, friends, and dad, with whom Nan was close, she had decided to relocate from New York to the San Francisco Bay Area. The move had been a while in coming, since Nan had often aspired to make her home here, where her sister Bess had lived for over thirteen years, and where Margot had settled after she and Nan had finished graduate school seven years ago. While the Bay Area didn't offer as lively a performance scene as New York, for the very same reason the competition as a music teacher was less steep. The coastal hills afforded phenomenal cycling opportunities. Further, Lilian had recently retired from Minneapolis to her vacation home in Marin County, and the thought of being near her again made Nan's heart leap with joy. Lilian felt like one more link to Nan's mom Gwen, who had been Lilian's close friend.

Nan had moved in early December 2019, full of hope for a new start and better opportunities. In mid-March of the following year, the pandemic closed everything down and effectively muzzled the music scene. Had her dad Frank not generously given Nan the money to cover her rent (which he insisted was not a loan but a gift), the troubled days that followed would have left her without a home.

So began a more desperate phase of Nan's perpetual battle to scrape together enough money to make ends meet while keeping healthy and fit and continuing to pursue all of her passions. Nan refused to compromise on food, exercise, music, art, or literature. Her mom had once told her she resembled Freud's *Schnorrer*, who was apparently a Jewish beggar. This *Schnorrer* had asked a wealthy man for money so that he could go to the most expensive bathing resort in the region to restore his health. The rich man had agreed to give him money, but asked why it needed to be that particular resort—why not some place cheaper? The beggar had replied, "Nothing is too expensive for my health." Nan preferred to identify with Jane Austen's women: well equipped to meet Georgian society's demands of feminine accomplishment as described by any number of judicious male leads who were eager to support those accomplishments.

Nan reached the cold and gusty intersection of the fire trail with Grizzly Peak Boulevard and turned around to descend, still pondering how to balance the practical elements of life with her pursuit of excellence. Whereas people often spoke of the elusive work-life balance, Nan personally puzzled over the competing pressures of survival and her own high standards of artistic, athletic, and intellectual achievement. She sometimes thought other people had been issued a secret manual containing directions for how to assign their time to different demands throughout the day and over the course of their life. Why had no such manual been given to her? She would have made good use of it.

As Nan flew down the last part of the fire trail and back out to the busy Centennial Road, she pushed away thoughts of what she lacked and tried to take stock of all she did have. Her one-bedroom apartment, though small, was not cheap—it was located in Rockridge, a little over a mile from the UC Berkeley campus—but its ready access to everything Nan needed more than made up for its higher cost. Since she was committed to not owning a car, being near the BART and dedicated bike routes meant safe and efficient means of travel. She had found that the Rockridge demographic was heavily musical, and all of her students lived within a five-mile radius of her place. Moreover, within twenty minutes, she could be cycling in the Berkeley and

Oakland hills with their lush redwoods and eucalyptus and their stunning views of the bay. Bess lived a mile away, towards the campus. Since Margot lived in the City and she got around solely by bike and public transport, she and Nan rarely saw one another more than once in a few weeks. But Bess, Bertie, and Pooh-bah had largely gotten Nan through the hardest part of the pandemic—first by including her in their "pod," and then by making themselves constantly available for physical contact, love, and affection. Nan knew she would never have survived the loneliness and depression that chronically threatened her, had she not had these three nearby.

After finishing her run and showering, Nan stopped by Bess's apartment and, giving a brief knock, opened the always-unlocked door. Dark-haired, serious Bertie was sitting on the floor surrounded by large pieces of a puzzle that featured a steam train. When she entered, he looked up and said, as if continuing an ongoing conversation, "I started with the caboose because it has more parts to it."

"That's usually the best way to go," Nan said, leaning down to kiss him. "Wow, just a few months ago, I remember you were working on a 100-piece puzzle. This one is . . . 300 pieces." For a few moments, Nan admired his concentrated, methodical way of working out the shapes, and his quick eye that compared the varied shades of color among similar pieces.

Meanwhile, Pooh-bah affectionately placed his fore-paws on Nan's knee and licked her face by way of greeting. With his long blond neck and back, he looked like a stretch-Chihuahua with some of the telltale facial features of the Chihuahua (like the big, alert ears and well-defined brow), but with a dachshund's soulful eyes and stubby, turned-out front feet. Bertie had selected him from among hundreds of dogs he and Bess had seen one spring day two and a half years ago while visiting four different shelters. Finally, at the Pleasanton shelter, across a crowded room, Bertie's eyes had singled out the erect ears and inquiring look of a Chihuahua blend who, when allowed out of the cage to meet him, had promptly rolled onto his back, closed his eyes ecstatically, and offered his arched chest and belly for Bertie to rub. It was love at first sight, and they had adopted the dog that day. When they'd sent Nan the pictures of him over the phone, she had suggested they name him Pooh-bah, fan as she was of Gilbert and Sullivan's operetta *The Mikado*. The name had stuck with Bertie.

"Hey, Nan!" Bess called out from her inner office. Bess worked as a statistical analyst for a non-profit environmental research institute based out of southern California that had opened a location in Mountain View a decade ago. Since the beginning of the pandemic, she worked from home three days a week, a change she welcomed to ease her commute and to get a bit more sleep. "Are you loving up that dog again? I think you really come over just to see Pooh-bah."

"Like Bertie in the nephew department, he sets the bar high for furry friends," Nan conceded. "What are you working on?"

"The same presentation I was working on last week," said Bess. "It's taking extra long because we're using this new software that one of our engineers developed this past summer. It's actually sort of fun, in a tortured kind of way."

"Does it take longer if you have to Zoom with all the engineers?"

"Actually, for most things, I think we save time remotely. Though it'll be nice to deliver the actual presentation in person on Monday. You and I will have to celebrate after that. Whom did you teach this week?"

"I have a new student, named Melanie, who's very sharp." Nan produced a banana for Bertie and one for herself, offering one to Bess as she joined them in the living room.

Bess shook her head. "What I need is just a small dose of caffeine." They hugged, and Bess went into the kitchen to make tea. "Any other students?"

"Monday afternoon I went to a house around the corner where those sixteen-year-old fraternal twins, Gerald and Jessica, live."

"Do they get a lesson together, like one of those oysters you open up and find you've gotten two for the price of one?"

Nan laughed. "No, but it is convenient, because Gerald usually has his lesson directly after Jessica, and they have similar learning styles. I can't help it, Bess, but I always prefer teaching the adults and young adults to the much younger kids."

"It's probably because you used to teach college-aged kids," Bess remarked.

"Yeah, I miss them."

"I know you do. You'll return to college teaching eventually—all these piano lessons are only temporary, until you can get your quintet composed and performed. Then you'll have your pick of academic appointments—*and* you'll be an even more adept teacher."

"But I never write!"

"That's only because right now, like tons of other people, you're struggling for survival. It stands to reason you don't feel inspired or motivated to do creative work that doesn't pay up front," said the ever-practical Bess. "Sometimes you have to step back a little in order to leap forward."

Bertie interjected, "Mom, can I do games on my tablet now? I've done part of the puzzle, like you said."

"Yeah, but only until the clock says 5:30. Then we're having dinner. And you have to put away the unused pieces into the box."

Bertie nodded, switching modes from material puzzles to electronic games with the alacrity of a quick-change artist. Pooh-bah inserted himself into the new activity almost as promptly, pushing his head up through Bertie's arms and delivering effusive kisses to the boy's face while Bertie delightedly stroked his forehead.

"Bertie is so sweet with Pooh-bah," said Nan. "I'm ninety percent certain that something Bertie does in the future is going to involve animals."

Bertie was indeed often proposing animal guessing games and revealing his fund of knowledge about all things Animalia. This past summer Nan had enjoyed a fun trip with Bertie to the Oakland Zoo, where he'd shown an intuitive understanding of classifications, history, and natural habitat. He had a gentle knack with every animal he encountered, right down to the beetles and butterflies.

As Bess made her tea, Nan continued telling her about her weekly student line-up. "Sunday afternoon I also taught fourteen-year-old Harley, who, I think I mentioned to you, is on the spectrum. He seems like a perfect example of how music can be a kind of therapy for kids with autism. He's working hard on those two Burgmüller pieces, and when I asked him—as you know I always ask students—what he's thinking of when he plays the "Arabesque" one, he said it reminds him of a squirrel hiding away food for the winter in a tree."

"That's a delightful image!" said Bess. "What about your chamber group? Are you still meeting regularly?"

"Jules, Céline, and I are working on a Beethoven trio, and after the holidays we'll all be preparing the Schumann opus 44 piano quintet. I'm really excited about that."

"And for both of those you'll be performing in the same space as when you played the Brahms piano quartet this summer? That was cozy."

"Yes, Jules has his house for the trio, and we're hoping Despina can talk her friend into letting us use his private concert hall for the Schumann—supposedly it has really good acoustics, and it has the feel of a large living room, so it's also somewhat intimate. It's in North Berkeley, though, so virtually every group in the East Bay wants to play there."

"Will you record any of these concerts?"

"Yes, Simon's doing that for us," said Nan. "He's a wizard at balancing sound."

"So you see, you're going to have two more recordings and performances to put on your résumé. No more worrying about what you're *not* doing—think only of what you *are* doing."

Nan laughed. "This is the real reason why I come over here so often."

Chapter Two

It was an unexpected treat to be able to pay a visit to her sister that Friday. Nan usually taught two students during this time slot. Today both students had requested earlier times in the week and she'd been able to accommodate them. Nan took a walk with Bertie and Pooh-bah around the block before heading west on her e-bike. Before Gwen had died, she and Nan had discussed how Nan could do volunteer work like tutoring kids in music after school, developing a program for music appreciation at a local community center, and accompanying musicians who couldn't afford to hire pianists. Nan agreed with her mom that music should be solidly based in community, and that by educating as many people as possible in music, she would be performing a needful public service. As a modest start, since the vaccine allowed her once more to share her music live with elderly folks, Nan had been biking over to West Berkeley, every Friday evening, to play piano at a retirement home where a diverse group of residents welcomed her by turning down the television in the common room and turning up their hearing aids. "There is no more forgiving audience than a slightly deaf one," Nan thought happily, as she settled onto the bench and opened up a book of Gershwin pieces.

Much like the meditative feeling of painting or cycling, the calm but intensely focused feeling that descended on her while she played piano allowed her to process thoughts in a free-form but productive way. These thoughts ranged from long-term dreams and plans to short-term problems related to teaching or composing. Today, while Nan played through an abridged version of Gershwin's "Concerto in F," it occurred to her—not for the first time—that if many of life's most important decisions had come to her while playing music, music itself could be potentially revolutionary. Having an audience while conceiving these ideas made them all the more real and potent. Nan wondered what Plato's *Republic* would've had to say about

Gershwin's influence on the public spirit. Socrates probably would've deemed it a baleful one.

So many times, when Nan had been manic, she had been certain that the music she played or composed, or the poetry she wrote, constituted a revolution in itself, an incendiary document holding the transcript for positive social change. By transmuting her suffering into art, she felt she was enacting, on a small scale, a public improvement. Backing up her assurance was the fact that she was never disappointed in the aftermath, when with cooler examination she turned over her compositions, recordings, or words and assessed their quality. She may have upset a few people in the course of her various episodes, but she always apologized with the sincerest humility and contrition. And she always viewed these episodes, in hindsight, as opportunities for growth.

From Gershwin, Nan moved on to playing some of her favorite pieces—Schubert Impromptus, Brahms Hungarian Dances, and Debussy's *Pour le piano*. To please a curly-haired resident of Irish descent named Maggie—knowing these were her personal favorites—Nan tagged on some Irish folk tunes towards the end. As Maggie and others applauded, Ethan, a dapper Black man who always dressed to the nines, requested some boogie woogie. Though not a natural improviser, Nan could muddle her way through a few barrelhouse blues and boogie woogie songs whose chord progressions she had built on and gotten to know well enough over the years to be able to rely on these as performance pieces.

"Will you be available to play for us the day before Thanksgiving?" asked Fannie, the energetic, seventy-something event organizer.

"I should be," said Nan. "Any special music I should prepare or bring?"

"It'll be a holiday high tea, so maybe something relaxing—like Chopin?" Fannie suggested. "But we're not too particular, as you know, hon."

"I look forward to it." Nan kissed Fannie on the cheek and called out goodbyes to everyone else.

Nan was grateful for the e-bike that awaited her outside. She had decided to cook a Japanese-inspired buckwheat-noodle stir-fry for dinner, and Berkeley Bowl, just down the street, had all the ingredients. The only problem with this paradise for healthy food-lovers, however, was the difficulties it presented in curbing one's enthusiasm: Nan often walked out of the store with twice as many bags as she had planned on filling.

Despite arriving less than a half-hour before closing time, Nan found Berkeley Bowl packed with people. As usual, it was hard to remember the precise items she needed as the wealth and variety of foods assailed her senses. She couldn't help leisurely smelling her way through the herbs and greens section, where three different kinds of arugula neighbored an array of parsleys and cilantros, and where she had to cave and treat herself to some much-loved savory watercress. After loading up a hand basket with the ingredients for the stir-fry, two bottles of the Bowl's current red wine special, and some bars of single-origin dark chocolate, Nan tried to put blinders on as she passed the ever-tempting cheese section. Cheeses changed the check total dramatically, she knew, and they always dictated that one buy bread.

But across from the cheeses were the warm roasted chickens and, hungry as she now was, Nan thought how she could snack on the drumstick while cooking, and then could put the breast from the chicken into the stir-fry. Using one hand to carry the chicken and the other to carry the basket, she proceeded to the check-out.

Outside, one of the young men from the local group Hip-Hop For Change engaged her in conversation. When he heard she had already contributed to the cause a few months before, he thanked her, making a peace sign. The air was chill with the weight of moisture, and Nan shivered as she loaded up her e-bike. Within ten minutes she had reached her place and parked her bike beneath the overhang in the driveway out back. As she came in the door, her phone rang. It was Thomas, her cycling partner. They had met while riding up the hill on a bracing early December day last year.

Never one to beat around the bush, Thomas opened with "What time are we meeting tomorrow?"

"How about 10 at the base of Tunnel? Shall we go to the bottom of Redwood?" Nan began to unload the bags while they talked, but she halted as she took out the chicken container and had a chance to examine it closely. "Uhh, Thomas, something has happened to my roast chicken."

"Yeah?"

"Yeah. It's missing both its drumsticks!"

Thomas burst out laughing. "Someone else was even hungrier than you."

"Do you think it was a homeless person?"

"Who else? Well, consider this your charitable act of the night."

"Though it's kind of gross to think they had their hands on this chicken," Nan mused.

"If you cook it up in your stir-fry, it won't make a difference," said Thomas. "Let this be a lesson to you: from now on, check your chicken. Listen, I've got to go. The bar is hopping now, we're short-staffed tonight, and we just got a party of ten in."

"See you tomorrow."

Nan tried to visualize Thomas right now at the bar he owned at 24th and San Pablo in Uptown Oakland, The Tippler's Haunt. Thomas was stupendously fit and, whenever she rode with him, he pushed her to her climbing limits. He had a dark sense of humor and spoke German fluently. Gregarious and dynamic, he also had an introverted side that somehow made him the ideal owner of a business where people came seeking not just a pick-me-up but a confidant.

Nan wondered, not for the first time, if Thomas had ever considered taking their cycling friendship to a more intimate level. She suspected he was single, but she was loath to risk what they'd had for almost a year now—a good pairing in power output and endurance on the bike, and similar M.O's while riding. Nan had a feeling the attraction might be mutual, and much of their conversation was flirtatious, but she balked at making the first move.

Margot had once suggested that they show up unannounced at Thomas's bar and now Nan wondered if they might not put this plan in action after all. She turned on KDFC, the Bay Area's classical radio station, and called Margot while chopping the vegetables for the stir-fry.

"You still at the gallery?" Nan asked. Margot had worked incredibly hard the last seven years to open up her own gallery in the Design District, a place that housed contemporary art. Late this past summer, finally, it had opened to immediate success. Currently, it offered an exhibit of modern choreographers' work displayed through video, drawings, costumes, objects, and painting. As Margot had pledged to do through all her exhibits, she featured the work of several local artists.

"Just got home," said Margot. "Romare barfed all over the bathroom floor."

"Ohh, poor thing!" Nan cooed. "Was it something he ate?"

"Like as not. He goes out on these hunts periodically, and I can't tell how much of what he finds he partakes of," said Margot drily. "Yes, Romare, we're talking about you."

"Do you still want to go to Thomas's bar some night?"

"How about tomorrow night?" Margot jumped at her suggestion. "I need to get out of the City and chill in the East Bay."

"Okay, only, should I mention it to him tomorrow while we're riding?"

"No! Let it be a surprise."

"I'm kind of nervous."

"What's the worst that can happen? I'll be there."

"That's what I'm worried about," Nan admitted.

♪ ♪ ♪ ♪

After a cold but invigorating forty-mile hill ride with Thomas, Nan got in three hours of piano practice. Then, after eating some soup, she considered what she should wear for their stealth crashing of Thomas's bar. The Bay Area never failed to put the kibosh on any attempt to bare skin, since after 5 p.m. temperatures dropped to wintry levels. Going out, one always had to balance the desire to look sexy with the necessity of base layers. Nan settled on pale blue, low-rise boot-cut jeans, heels, and a clingy black long-sleeved, long-waisted top that she could layer with a close-fitting leather jacket and scarf and, over these, while in transit, a fleece coat. Nan never wore make-up—she was super sensitive to putting anything on her skin other than certain moisturizers, and disliked the smells of most products—but she loved choosing earrings and necklaces to dress up different occasions. Tonight she selected gold hoop earrings as a simple but effective complement to the rest of her outfit.

Every time she picked out clothing from her closets, she was reminded of the time eight years ago that she had gone out and, in the course of several days' shopping, spent ten thousand dollars. Steve, the man she had been dating for two months and with whom she had been falling in love, had left a note on her piano keys that read, "You want to live in a city ultimately, and I want to live in a rural area. You want to have kids, and I don't. Wishing you the best." Nan was still paying off the debts from that manic spree, and still trying to make the fullest use of all the clothes she had bought.

Once arrayed for the night, she rode her e-bike to meet Margot at the corner near Thomas's bar. Margot had pulled her smooth jet-black hair back in a casual but elegant chignon and she wore a bright-gold suede mini-skirt that highlighted her honey-toned skin. To meet the cold, she wore white silk stockings underneath.

"You look gorgeous," Nan said.

"So do you." Margot's eyes sparkled.

After they'd exchanged a long bear hug with one another, Nan locked up the bike.

"Shall we?" Margot linked her arm in Nan's and directed them towards the bar.

"You took the train to 19th street?"

"Yeah, but I may Uber home, depending on how late we stay." Margot winked at her.

When they pushed open the door to the bar, the warmth of dozens of bodies and the swell of sound from voices raised to match the Brazilian music that played overhead zapped Nan with a fresh buzz of excitement. Here goes nothing, she thought, as they approached the bar.

She caught Thomas's eye and noted that he seemed outwardly unfazed by their arrival. Either he considered it the most natural thing in the world that they should be here right now, or he was an excellent actor. Nan smiled shyly at him, and turned over the drinks menu that one of the bartenders had handed her. Margot followed her gaze and then faced back around to Nan, raising her eyebrows expressively.

"Does this constitute an invasion of a person's territory?" Nan asked Margot in a low voice.

"If you're asking, is this a bold move, then yes," Margot replied. "But from the way you've described this guy, he won't be turned off by bold moves."

"Can I get you ladies anything special?" Thomas had a soft baritone voice, and Nan suddenly realized his presence was magnetic.

It was fortunate, since Nan's tongue was tied, that Margot stepped in to fill the gap. "What would you recommend for two cocktail lovers?"

Nan regained enough composure to mutter, "Thomas, this is Margot. Margot, Thomas."

Thomas nodded at Margot. "A pleasure. It depends what taste you're looking for. We have sweet, bitter, sour, spicy, smoky, and various combinations of those."

Margot considered this. "Spicy and sour sounds nice."

"Our 'One-Two Punch' is served up without beer, if you like something strong made with Scotch."

"What about you, Nan?" Margot asked.

"Oh, um . . ." Nan cast a quick glance over the double-columned list of drinks, finding the words a blur. "Do you have anything bitter and smoky?"

"I have something that's not on the menu that I crafted myself with mezcal, Punt e Mes, and Amaro Nonino." Thomas's eyes glittered with a hint of challenge.

Nan managed a quick nod. "That sounds perfect."

When he had stepped away, Margot fanned herself a bit. "Is it hot in here, or is it just me?"

"He's hot, right?"

"It's not just him, it's you two interacting. Man, now *there's* some sparks."

Nan blushed deeply. "You think he's interested?"

"You can't tell? He leaned in to tell you about that drink, and even when he was looking at me he was so obviously aware of your every move."

Embarrassed, Nan tried to reclaim some distance. "I like the way when we ride he's competitive in a nice way—it's as if we're both working together to hammer it. And he's super smart. He studied in Austria for a year and knows French as well."

They paused their conversation for a moment as Thomas came in view again and began to make their cocktails. Nan realized for the first time that a large part of the pleasure of a cocktail was enjoying watching someone make it. Thomas had rolled up his sleeves and Nan couldn't help but zero in on the well-defined vacillation of his strong forearms as he prepped the drinks.

Nan changed the subject. "Can I come see the exhibit some time after this next week?"

"Are you free on a Wednesday afternoon? That's a slower day, generally, and I'll be able to show you around more."

"I can't wait. This one seems to have made even a bigger splash than your food-themed exhibit that opened the gallery this summer," Nan enthused. "What's next after this?"

"I'm thinking of something Taiwanese and Taiwanese-American next," Margot mused. "It'd be heavily historical. There are plenty of California artists to showcase, Taiwanese identity is a hot topic now, and my parents would be super pleased."

"Would they come over from Taiwan for it, do you think?"

"Definitely. Since I never got the chance to visit them this summer because of the gallery's opening, we're all keener than ever to see each other again for the first time since before the pandemic."

Thomas poured their drinks and garnished them in a few deft movements. "Let me know what you think."

Nan took a sip of hers and couldn't keep from emitting a low moan.

Margot laughed. "I guess that means it's good." She tasted her drink and promptly raved, "Wow. That really does pack a punch. It's got a nice lavendery aftertaste."

They swapped drinks and exclaimed over the startling tastes, so different from one another and yet, when combined, so harmonious.

"What do you call this one?" Nan asked, palming her own drink.

"I'm thinking of calling it 'Giving Best,'" said Thomas. When he saw their questioning looks, he added, "It's what the hunters say of the fox when the fox successfully evades their chase."

♫ ♫ ♫ ♫

Though Thomas was too busy for the rest of that night to speak much with them, Margot and Nan stayed a good while to soak in the atmosphere of the bar. The music shifted from Brazilian to Cuban, and some people started dancing with verve and panache. A group of college-aged kids arrived around eleven—probably when some other bar had closed and kicked them out—and they ordered impossible quantities of beer. Nan saw the bouncer scrutinize each of their IDs before grimly nodding them in.

"Why *is* the Bay Area such an early place?" Nan voiced a thought she'd had numerous times. "In New York, this would be considered the beginning of most people's nights. Here, this bar is on the late-late side just because it stays open past 11."

Margot sighed. "I know, it doesn't make much sense. Some people have told me it's because the West Coast has to keep to an eastern schedule. Others have explained it as a family-centered trend—but that shouldn't apply to SF. And it doesn't quite sound right to say that it's all because the BART closes down at night, unlike the New York subway system. I mean, wouldn't BART change its times to suit its customers' needs?"

"It's like all the stores and restaurants out here that close Monday and Tuesday or stay open only 12 to 6. They could make a killing if they just stayed open two more hours or if they only closed one day a week. In New York, some employee would always be eager to hold down an extra shift or work an extra day, and a New York business is never going to turn away customers or resist the chance to make more money.”

“Yeah, here it seems to be a quality-of-life decision that the entire culture has embraced. I kind of respect them for it,” said Margot. “But it is disappointing when you need something outside open hours. And it takes some time adjusting to the fact that sushi, pizza, and bakery items just aren’t options on Mondays or Tuesdays.”

“I wonder if the pandemic is going to accentuate that trend from now on. ‘The Great Resignation’ seems to have everyone rethinking the meaning and value of their occupation anyway. I imagine cutting some more hours could be one way employers sweeten the deal for employees they want to retain.”

“This could be a real blow to capitalism,” Margot suggested. “A return to simpler times when community and tradition outweighed greed.”

“Or not.” Nan matched Margot’s sly smile, taking a last sip of her cocktail. “Have you met anyone special recently through Select Singles?”

“I’ve had a bit of a hiatus since things heated up with this exhibit,” Margot confessed. “But I have been texting off and on for a while with that British guy, Aaron. He’s got a wicked sense of humor and he seems super cute from his pictures.”

“When are you guys meeting up?”

“His work got busy right around when mine did, so we agreed to wait. But I don’t want the flirty momentum to subside, so I suggested we get together Thursday night. He works in the City, so it won’t be hard for him.”

They settled up the tab, finding that they’d been charged only for their last round of drinks. Just as they were putting on their coats, Thomas appeared at their side.

“Now you know some of the craziness that goes on here,” he said. “Sorry I couldn’t talk more.”

“Thanks for some really delicious drinks,” Nan said. “Are you up for a ride on Wednesday midday?”

Was it her imagination, or did Thomas get closer into her personal space as he replied? “Sure. This time let’s go over Grizzly.”

They both leaned in and kissed one another’s cheek, with a split-second of lingering beyond the normal. “See you then.”

Nan's heart rate had considerably sped up as she and Margot walked out the door.

"Yep, the coming week is suddenly looking pretty dynamite." Margot gaily hold of Nan's arm.

"Way to make a bike ride look tempting. By the way," Nan said, turning to Margot. "Who is the fox and who the hunter here?"

"I think you're both meant to be both," Margot retorted diplomatically.

♪ ♪ ♪ ♪

"Teddy's getting married?!" Nan nearly dropped the phone when her dad made the announcement, which Nan wasn't sure she'd heard correctly.

"He's known Naomi for over two years now," Frank pointed out, "and they've saved up enough for a small house over on Grand and 47th street."

"Wow." Nan had so many questions, she didn't know where to begin. "Where? And when?"

"For both the ceremony and the reception, they've booked The Blaisedell for April 2. The wedding will be a little bit smaller, of course, given the pandemic—only about 150 people."

"Teddy even *knows* 150 people?"

"Well," Frank admitted, "a lot of them are on Naomi's side. And you know her family comes from Chicago, so people will be flying and driving in for the weekend. But you all will be coming from California, and then there's my brother's kids from Washington—if they can come—and Gwen's sister Jo's three and their families from Florida and Louisiana. So let's just hope travel isn't too hard at that time. You never know, with the ups and downs of this virus."

"But Teddy's so young!"

"He's 31. That's the age I was when I married your mom—who was only 25."

Nan was still reeling from hearing of Teddy's promotion, a few months ago, to Head Game Developer at a studio in Minneapolis. She recalled his endless struggles finishing his bachelor's in math at Macalester and, following that, the doubts everyone in the family had that her brother would ever be employed, given his lifelong penchant for puzzles and resistance to work. He had lived with Frank and Gwen for a few difficult years when he'd held only the desultory job as a car

26

wash attendant, a delivery person for Instacart, or a chess instructor—the latter a job he had, unfortunately, loathed. Unlike Frank, Gwen, and Nan, Teddy was no teacher. He had surprised them all by teaching himself coding four years ago and by building a successful gaming app for the ever-popular game Crossfire. Meanwhile, he had steadily rolled out samples of his work on blogs and discussion boards, constructing a network and garnering interest from game developers.

Upon being hired at his current studio, OneUp, Teddy had met Naomi, who was a game designer with a bachelor's in psychology. She not only oversaw level design, but helped brainstorm the characters, plots, and themes from each game's conception. While Teddy never said anything outright, his face declared his admiration for Naomi's talents whenever she was near.

"Isn't Naomi Teddy's first girlfriend ever?" asked Nan.

"The first to write home about, I think," Frank hazarded.

"How did he propose?"

Frank chuckled. "Oh, nothing fancy there. You know Teddy. As far as I understand, he took her out to brunch at Revival—over on Nicollet. When her eggs Benedict arrived, he passed her the hot sauce, and around the neck of the bottle he'd placed the ring. She nearly missed it though!" Frank chortled again to think of the scene.

"Wow, everything's really coming together for Teddy. I'm happy for him." Though Nan spoke sincerely, her voice contained a quaver that Frank picked up on immediately.

"You know, everyone has a different journey to finding someone who's right for them. Yours and Bess's may be a little longer or more involved, but the outcome will make it no less worth the wait."

"Thanks, Dad. I admit Teddy was kind of my last remaining proof that it was something family-related about all three of us that kept us from pairing up as normal thirty-somethings do. Now, it looks as if the pressure's on me and Bess to prove ourselves capable of meeting someone."

"Darling, this is one area where you really can't compare yourself to other people, tempting as it is to do so," Frank urged gently. "Just keep living your full life and developing your pursuits as you do, and it will happen naturally."

"Sorry, I didn't mean to take the spotlight away from Teddy. What sorts of events are they planning around the wedding?"

"I'm having the Rivkins here for the whole week, which I'm looking forward to. I've never met them. Did you know Naomi's

Father Efraim is a retired master builder who worked on the James R. Thompson Center?"

Frank himself, who was 72, looked to be teaching mechanical engineering at the University of Minnesota until he keeled over at a ripe old age, as Nan hoped. He loved teaching, loved his students, and loved research, in that order. He lived in the same house, in South Minneapolis, in which Bess, Nan, and Teddy had been born. Nan knew Frank was keen to have grandchildren run around the house and fill it with young sounds.

"I had no idea. Will the ceremony be religious?" Nan asked.

"No, neither Teddy nor Naomi wanted that," said Frank. "Just an all-purpose exchanging of vows—no bridesmaids or groomsmen even."

Frank came from a long line of Episcopalians in northern California. Gwen had been raised Quaker by a family who came, on both sides, from Quakertown, Pennsylvania. She had, in turn, raised Bess and Nan to follow Quaker beliefs and to attend meetings. Nan's attendance had fallen off since the pandemic, and she had come to consider herself more of a secular humanist. Of all three siblings, Bess was probably the most religious, as she occasionally attended Episcopal church services. By contrast, Gwen's sister's family, living in the South, were Evangelical.

Nan volunteered, "Well, let me know how I can pitch in. You know I'm good for anything music-related."

"Teddy has some ideas about ways you can help," said Frank. "And he definitely wants you to play something at the reception."

"Okay, no problem. This'll be a fun occasion to compose a piece in his honor. Wait." Nan was struck with a thought. "Will we all be expected to bring plus-ones?"

"You certainly can." Frank's voice rang with enthusiasm. "Any special person in mind?"

"You never know," mused Nan. "A lot can happen in five months."

♪ ♪ ♪ ♪

Tuesday evening, as Nan practiced Beethoven's "Ghost" trio with two of her chamber cohort, she was glad to have the external confirmation of one of Beethoven's most dynamic trios to remind her of the fire within her. In the third movement, Jules's violin and Céline's cello kept up a lively dialogue with the piano that alternated between the keys of D-major and B-minor, and she felt as if she were literally soaring as she crescendoed chromatically to the final chords, which the three played with precision and a flourish. They all laughed in triumph, panting slightly.

"We should've played that on the street on Halloween," Céline declared. "That's about as haunted a piece as Beethoven wrote, I think."

"Yeah, that eerie interplay of major and minor keys in the last movement sounds unstable," Jules remarked, putting more resin on his bow.

"It's my favorite of all the trios," said Nan simply. "Especially the scary second movement."

"I heard it was Czerny who gave the piece the name 'Ghost' after he heard that movement." Céline took a sip of water. "But it wasn't the ghosts of *Macbeth* he was thinking of—rather, the ghost in *Hamlet*."

After they'd put away their instruments and stands, Jules invited them for a glass of wine. "No more deep concentration required!"

Nan loved communing with fellow musicians, especially her chamber group. As so often happened, this group found its roots in Meetup.com. It had taken some false starts and cautious approaches to various pieces with various permutations and combinations of players before a core group of five of them had solidified in early March 2020. These five well matched each other in ability, approach, and aims. Though the lockdown had briefly paused their efforts, in July 2020 they had been able to pick up where they left off, once they could open the back garden door leading out from Jules's family room, where his piano resided. The string players would arc themselves around just outside the doors, facing the piano, and all of them, masked and at a distance, would play as if their music could quell the pandemic.

"Have we decided to come up with a name?" asked Céline.

"My son Josh thinks we should call ourselves 'The Hellraisers.' Teens and horror flicks, you know." Jules smiled.

"Are we planning to use the same name and just attach it to 'Trio,' 'Quartet,' or 'Quintet,' according to the context?" Nan asked.

"That seems simpler," Céline submitted. "For those of us who are fans of astronomy, we might explore names like 'The Perseid Quartet' or 'The Orion Trio.'"

Jules and Nan nodded enthusiastically.

"Weren't the Perseid showers just a few months ago?" asked Jules. "We were camping in Yosemite—wow, was that spectacular."

So they decided to propose the name "Perseid" to Despina and Paul over text. They finished their wine, and Céline and Nan headed out.

"How are Max and Cécile doing?" Nan asked Céline as they stepped towards their respective rides.

Céline pouted. "Max is jealous of Cécile because she's heading off to college next year. These college applications are affecting the sanity of us all!" She laughed. "Arthur and I would love to have you over for dinner, by the way, maybe the weekend after next, if you're free?"

Nan beamed. "I'd love to come! Let me know what I can bring."

"It seems after next weekend it's nothing but one long string of holidays for a month and a half." Céline sighed. "I feel exhausted already, you know?"

"It'll go by in a flash, though—it always does."

"Well, in the midst of it all, I say, thank God for music."

"Thank God for music," Nan agreed fervently.

Chapter Three

That night Nan dreamed of Thomas. Flashes of him hovering over her, bare-chested with a predatory look in his eye, his black hair tousled and one strong arm pinning her arms above her head, combined with a feeling of the weight of his muscular legs on her body. When Nan woke, she threw off the weighted blanket she used to sleep more deeply and gulped down some of the water on her bedside table. What was Thomas thinking? What was his house like? Did he want her anything like the way she wanted him? How would their ride be different tomorrow after what happened Saturday night? Nan's thoughts raced, until she calmed them down with her meditation techniques. "Quiet the mind," she repeated, breathing deeply. She counted breaths by multiples of 13, until, somewhere around 455, she dropped back off to sleep.

In the midmorning sunshine the next day, she and Thomas greeted each other briefly on the hillside and started climbing up Tunnel Road. While they had often been chatty from the outset of their rides, today only their synchronous pedal strokes and nearness signaled that they rode together. They stopped at the top to put more layers on—Grizzly Peak was always gusty and five degrees colder than the climb up—and Nan had a frisson of excitement as she helped Thomas tuck in his buff beneath his skull cap. She briefly imagined him catching her hand and pulling her to him.

They started their climb over the peak.

Nan first broke their silence. "Your bar is pretty awesome. Where does the name come from?"

"I once heard someone read out a crossword puzzle clue that was 'tippler's haunt.' The solution ended up being 'bar.' I remembered it years later when I was renaming this place."

"How do you get to be the owner of a bar?"

They had slowed a bit from their usual pace, so that they could hear each other and converse more easily. Each time they detected the sound of a car behind them, they reverted back to single file and paused their talk.

"There are lots of ways it can happen. In my case, I had the money when the former owner was selling the place and I'd worked for him for a while, so he chose me over a few other contenders."

"How long ago was that?"

"Eleven years ago."

"Just curious, what was the place like previously? I mean, that area hasn't always been so savory."

"Yeah, it was a real dive back then. If I tell you its former name, you know everything you need to know."

Nan looked over at him expectantly.

Thomas smiled. "'Harry's.'"

"That's it? Just 'Harry's'?"

"Not very original, I know."

"I can just imagine the kind of clientele Harry's attracted." Nan chuckled. "Probably the opposite of the kind of folks who come into The Tippler's Haunt."

"You'd be surprised at how many regulars are still there from the old days, though," said Thomas. "These people have been coming for decades and they'll probably still be coming a few decades hence. It's like an old habit—it dies hard."

Nan nodded. "Or an old home. I once read a really moving essay developing the idea that all animals and plants have not just a strong memory of home, but an overwhelming impulsion towards it at all times. The author's best example was of the Philadelphia El train getting abandoned when the subway was built. All the pigeons and a blind man who used to beg up there suddenly had nowhere to go. For a while they kept coming back to this long-standing source of food and shelter, expecting the life of the place to resume. Then, as soon as the demolition crews came to tear down the El tracks and platforms, the birds and the blind man came back once more, full of hope that this was going to be the return of their routine and livelihood. But the whole structure got dismantled and nothing went back to the way it was. I think the essayist compared the former bustle of the El to a 'food-bearing river.' Your watering-hole is probably like that still for some old regulars."

"Those same regulars would probably be baffled to know they're being compared to Philadelphia pigeons," Thomas remarked.

As they started the steep ascent up Grizzly, Nan asked, "Have you ever had to throw anyone out?"

From directly behind her he replied, "I hired a really good bouncer in Benny. He breaks up fights, throws problem people out, and vets everyone who comes in."

Once they had gained the plateau portion of the peak, she ventured, "Were you surprised to see us the other night?"

"I figured it was only a matter of time."

"What do you mean?" She was intrigued.

"Well, you told me about Margot before, and I know you like drinking," he reasoned. "You may have forgotten, but you also told me Saturdays and Wednesdays are your nights off."

She was momentarily struck mute contemplating how much he remembered from their conversations. "Are you one of those people with a mind like a steel trap?" she asked laughingly.

It was his turn to chuckle. She loved the sound of the low rumble. "In what I like to think matters."

"Do things matter because we've remembered them, or the reverse, do you think?"

He became quiet for a moment. "Maybe it's a case of a virtuous cycle. We remember things because they matter to us, and then they matter even more to us as we continue to relive the memories of them."

"I guess then, as we get older, it's not so much that we lose our memory as that the memories we have become more intensified."

"That was certainly the case with my grandfather," he reflected. "He was one of the best storytellers I ever knew. Somehow even when he was telling the same story for the fortieth time, it was still funny."

"Where was he from?"

"Hungary. His family settled in New Jersey in the late 19-teens."

"Is that where you're from?" There was almost nothing Nan loved talking so much as heritage—as long as people were willing to open up about it.

Thomas nodded. "A suburb of New Brunswick."

"Did you go to Rutgers?"

"No, Columbia. I had to have a little distance from my family."

"So is 'Gellert' a Hungarian name then? I'd assumed it was German."

"It's Hungarian, but has Germanic origins," he said. "I see you're no Harry Potter fan."

"Oh, is that a character in the books? I've never read them."

"I had to read some of them to my nephews and nieces. And to arm myself against all the flippant comments people make whenever they hear my last name."

"I'm sorry for that. Have you ever been to Hungary?"

"Twice—the first time when I was eight. I still remember my dad's third cousins serving up a cottage cheese dumpling for dessert and that was the first time I'd ever had it. When we got back home, I was asking my parents nonstop for this sweet from the Hungarian bakeries. It's called 'túrógombóc.'"

She marveled at his pronunciation, which sounded pretty accurate. "Wait, do you also speak Hungarian?"

"Just little bits that my grandparents taught me when I was growing up."

"You must've visited Hungary again when you were staying in Vienna?"

"That was my second time. Then I could actually appreciate the countryside where my distant relatives live. I want to go back after the pandemic and cycle around the entire country."

Since they had, by tacit consent, decided to climb the steep Miner Road from Camino Pablo in Orinda, they devoted their breath for awhile to the effort. Nan began to wonder if she was just incredibly slow to realize the potential for something more in what she and Thomas had. Perhaps all this time he had just been waiting for her to give the signal.

"Have you applied to any more lectureships out here?" he asked, after they'd reached a flat portion of route again.

Again, she appreciated how closely he had listened to every last detail she'd shared about her career goals and missteps.

"A few. None of the ones I've applied to this fall has replied back yet. Obviously I'd love to be considered for a lectureship at Berkeley."

"I don't see why they wouldn't want you," he countered.

"Nowadays lecturers are expected to have a more impressive set of publications on their CV's than I have. I have only a few recordings, some modest performance experience, and a brief article in an education journal."

"What about your compositions?"

"I haven't published any yet. I have lots of bits and pieces in my notebooks, but they need to be put together and fleshed out."

"What's keeping you from doing that?"

"Cycling, among other things," said Nan.

"What other things?"

"Poetry, singing, painting . . . bar-hopping." She hoped to strike a lighter note with the last one.

But Thomas remained in earnest. "Judging from your determination and dedication in riding, I feel pretty confident you could complete a major composition if you really set your mind to it."

"Why does a person keep putting off what she could—and should—do?"

"Perhaps because she's afraid of what might happen if she actually completes it."

Nan was taken aback. "Why would she be afraid?"

"Because beyond it lies the unknown."

"But a *good* unknown, surely?"

Thomas somehow managed to shrug with his entire torso while cycling, a feat Nan wouldn't have thought possible. "Sometimes a bad known can be preferable to a potentially better unknown. Isn't that something Hamlet says?"

"'And makes us rather bear those ills we have than fly to others that we know not of,'" she offered.

"Shakespeare always puts it best," said Thomas.

♪ ♪ ♪ ♪

"Would you like to get some food this evening?" he asked. They had finished their ride and stopped to say their goodbyes at the place where Tunnel Road split off to lead towards his place.

"That sounds great," said Nan. "In your neighborhood or mine?"

"I'm tired of my 'hood," said Thomas, who lived in the Grand-Lake area by Lake Merritt. "How about Rockridge? I can park at your end of College and we can walk up 'the strip' towards Broadway."

"What time?"

"Seven?"

"Perfect."

Nan was so excited she couldn't contain her elation as she sailed down Tunnel Road to the Uplands neighborhood on her bike. "Wheeee!" she cried, her voice dissolving in the wind and traffic.

The afternoon flew by as she deliberately tried to fill it with piano, stretching, and a visit to Bertie, Pooh-bah, and Bess.

"Remember, no one out there is a Jane Austen hero," cautioned Bess.

"What do you mean, exactly?" asked Nan, thinking she knew exactly what Bess meant.

"Everyone has not just general flaws, but things that are going to annoy you personally. You have to be ready to compromise."

"Is that why we're both still single?" Nan teased. "Because we're unwilling to compromise?"

"No doubt," said Bess grimly. "You know the older neighbor upstairs, Donald, who's often watched Bertie for me—he once said, 'You and your sister are wild, untamed, and free.' He has a point."

"When are we celebrating your finishing the presentation?" Nan adroitly changed the subject.

"How about Friday afternoon before you go to the retirement home? We can go to Sliver—my treat."

Hearing the word "Sliver," Bertie declared, "I want a cheese pizza."

"Cheese it is." Nan hugged him. "Friday actually works for me again this week. I'd better go. See you both in a couple of days!"

A few hours later, seeing Thomas's text that said he was in front of her place, Nan came out and watched him park his black BMW sedan in a tight spot a little up the street from her apartment house.

"How do you parallel park so precisely without any of the modern parking gimmicks like those beeping screens with the boxes?"

Thomas laughed modestly. "My dad taught me always to turn around fully and look behind you before you park, and to keep checking the whole time you park. It's an old-fashioned, but tried-and-true, technique."

"What year is this car?" For the first time Nan found herself turned on by a man's choice of car. From the front it looked a beast.

"It's a 2004 3 Series. Kind of old and only partly electronic, but I like the style better than the current BMWs out there." He bleeped the locks with his key.

"You've owned it all that time?"

"I bought it in 2009—off an old lady, if you can believe it. Now it's got more than 134,000 miles on it."

As they walked up the street together, their bodies were as close as any of the couples they saw, even though they weren't holding hands or linking arms. Nan wondered if they could manage this precariously subtle but steady pressure only because of the excellent balance they had from riding the bike. Thomas was about two inches taller than Nan's five-nine-and-a-half, a fact for which she was grateful; so often she towered above the men she dated.

They stopped to look over a few menus along the way, finally settling on a French bistro-style place that Thomas said had good steak. Nan was very happy with the bar seats they had, which allowed them not only to have their thighs touching, but to brush arms as they talked. By now she was heady with the tension developing between them. Flashbacks to the dream she'd had the night before continued to play in her mind's eye, but she pushed them to the back of her brain so as to allow the reality before her to play out.

As Thomas bent his head to examine the drinks menu, Nan studied his eyelashes surreptitiously. They were a lighter shade of brown than his black hair, but they perfectly accentuated his bold, brilliant hazel eyes.

"I'm going to guess which cocktail you'll choose." Thomas looked up from the menu.

"Oh, okay." Nan glanced at the menu for the first time; after a few moments, she said, "The Bald Eagle looks great."

Thomas nodded. "That's the one I thought you'd get."

"Why so?"

"You love mezcal and, even though you liked my drink the other night, I could tell you liked the sour pucker of Margot's even more. The Bald Eagle gives you the best of both worlds."

"I'm feeling at a sore disadvantage here in two ways," said Nan, laughing. "First off, you're an expert mixologist. Second, I've had no chance to find out what you like to drink. Can I at least try to guess which cocktail you gravitated towards?"

"Shoot."

Nan studied the menu again. "Okay, this is kind of a shot in the dark, but I'm going with the Horseradish-Black Pepper Gin Gibson. I'm banking on your liking spicy stuff." She looked up hopefully.

"That does look tempting, I'll admit," Thomas conceded. "You guessed right in one way: I do love spiciness. I think I'll go with that one."

"Which one had you chosen though?"

"I have a weak spot for anything licorice-tasting," he hinted.

"Oh! The Earl of Orange. That does look good," she murmured. "Can we get that on our second round?"

After they'd selected some appetizers and entrées and put in their order with the bartender, Nan decided to bypass the easier conversational routes and cut right to the gnarlier ones. After all, she figured, she and Thomas were not exactly weaklings, and this was far from their first sparring session.

"So, this is interesting for me," Nan began boldly. "I've seen your competitive side this entire year, but I've also seen how 'sporting' you are, as I think the Brits like to say. And these days, I'm seeing what a gentleman you are. If it came down to a contest between competiveness and gentlemanliness, which one would win out?"

Thomas looked her shrewdly in the eye, without flinching. "It depends entirely on the company I'm keeping."

"But if, as in this case, the company is the same female when riding and drinking, what then?"

"I believe this may be what's called a false dichotomy." He sipped his drink. "You're assuming that one or the other must apply in the two contexts you've named; whereas there may be another possibility."

"What might that be?"

"Suppose, for instance, that a third, self-interested motive governed both scenarios."

She was temporarily nonplussed. "How could the sociability of playing the part of a gentleman ever be self-interested?" Something struck her and she blushed and became silent.

Thomas took another sip of his cocktail, holding her gaze. "Human motives are often more complex than they appear at first glance."

Nan rallied. "Okay, Monsieur Poirot. Point taken. It's easy to oversimplify the grounds of human behavior. But somehow I feel you've dodged my original conundrum."

"It seems what you're really asking is, whether the truth outweighs tact, or vice versa, in my worldview."

"That's exactly it." She nodded.

"Maybe you had the same moral philosophy lesson in undergraduate days that I had . . . I think the lesson went, 'When you're considering telling someone a truth, make sure (1) that it's the right time and place to tell them this truth; (2) that, to the best of your knowledge, the truth you're telling them is accurate; and (3) that your motives for telling them this truth are legitimate.' I personally find that under these conditions, it's impossible ever to speak at all."

"Which one of these conditions is the hardest to fulfill, do you think?"

Thomas laughed. "Any one of them falls apart when submitted to the cold, harsh realities of the everyday world. Just consider a pro-vaxxer speaking to an anti-vaxxer. It may be the right time and place, from the pro-vaxxer's standpoint, to advocate getting vaccinated—after all, Covid death rates are much higher in red states nowadays—but to the anti-vaxxer, the timing seems suspect, given that the pro-vaxxers are in power. Or take the very different ways these groups approach the world and the question of 'accuracy.' To the pro-vaxxer, science represents accuracy; to the anti-vaxxer, accuracy is measured by more traditional values like religion. And we don't even need to get started on the question of the pro-vaxxer's motives here—the anti-vaxxer is sure to doubt those from the outset."

"Okay, so you've established that in many social contexts the truth requires tact in order fully to be truth. I agree." Nan sipped her drink. "But surely that doesn't apply when you're competing with someone? The true winners of the New York marathon the other day were Kenyans. They didn't have to worry about the question of tact when they crossed the finish line first."

"But is competition at the heart of what you and I do every Wednesday and Saturday? Or what we're doing now?"

Nan considered this. "It's certainly an important component—and a large part of the fun. I think in the medieval European courts, the playfulness of competition even took precedence over the question of objective justice. Whoever won the trial by combat had God on their side and therefore sustained the just cause. For that matter, in today's courts, the jury often goes with the side that seems to have provided better entertainment."

"I'm not sure if you realize it, but you've just reversed the direction your argument was taking." Thomas smiled. "If a definition of 'tact' is 'judgement' or 'prudence,' then you've just said that, historically, competition *was* tact."

"Yes, I suppose this drink has already muddled my head," she said, conceding the point.

"I don't believe that for a moment. But ultimately, to take your original question in the sense in which I believe you intended it—" He tilted his head at her meaningfully—"I am less swayed by the conventions society dictates, and more by what I believe to be right or true. So for me, truth always wins out over tact."

"I suppose I should take that as a compliment then," she observed. "You haven't been holding out on me when we've been riding together."

"Far from it." He swallowed the last of his drink. "You've got genuine power."

As the bartender set their appetizers before them, Nan admired how Thomas smoothly took the opportunity to put in an order of one Earl of Orange and another cocktail whose name she couldn't catch. She didn't say anything, but waited in pleasant anticipation for what would come.

"Is it only because you're a cyclist that you don't judge me for drinking too much?" she asked playfully.

"Are you suggesting cyclists are all heavy drinkers, and it takes one to know one?" he volleyed back with a smile.

"No, but surely you've noticed the trend in the last twenty years or so, where people are called alcoholics if they enjoy more than one or two drinks a week—and where women particularly are shamed if they drink at all with relish."

"I just ascribe it to the same kind of ignorance that looks at cycling as a poor man's form of transportation." He dipped a truffle-oil cod cake in the aioli sauce and took a bite.

"But doctors are in on it too," she said with feeling. "If I confess to my doctor that I enjoy more than one drink a week, I won't get any Ambien."

"That's just the party line they have to feed their patients," he said dismissively. "I've known plenty of doctors in my day with real drinking problems."

"Does this stigma go back to the Puritan founding fathers, do you think?"

"No doubt. I grew up with strict blue laws in the Tri-State area—and you probably had something similar in Minnesota."

"What gets me is that society assumes each person's tolerance is identical," she said, airing a longstanding grievance. "I'll bet anything that you and I have a much higher tolerance based on our fitness levels alone."

"Metabolism does have a lot to do with it," he admitted. "But imagine if you and I could make the argument—to the self-righteous finger-pointers—that our respective cultures value alcohol consumption, and that we're engaging in an important cultural tradition. On that basis, it might be discriminatory for doctors to deny us Ambien or for our blood alcohol content levels to be admitted as evidence in court."

"So far as I can tell, Hungarians and Scots do have heavy drinking in common." Nan unabashedly raised her glass to Thomas's. "By the way, what is this second drink?"

"I ordered us a Black Bottom Stomp, since I wasn't familiar with it, and the ingredients looked intriguing."

Nan examined the menu, which described the Black Bottom Stomp as a riff on the Black Bottom: cognac, blackberry brandy, cointreau, gin, lemon juice, orange juice, and, in this case, spicy syrup.

"Mmm, it's delicious," she hummed.

"Back to the subject of higher fitness levels." Thomas smiled roguishly. "What motivates you to punish yourself as you do?"

Nan nearly choked on her bite of salad. Then she realized he must be talking about exercise. "Oh, you mean in resistance workouts and runs?"

"And even in cycling," he persisted.

"Well, to me, cycling has so many benefits beyond fitness that it doesn't count in the area of punishment." She smiled crookedly. "But it's true that when I do my resistance workouts, or when I trail run, these all feel very much like controlled means-towards-an-end types of activities."

He leaned in towards her, and goosebumps rose on her arms in response. "So you wouldn't call what we did today a punishment as well as a reward? We did climb more than 4,300 feet in elevation, and rode almost 40 miles."

She fought to preserve her presence of mind. "Okay, I'll admit I have as much masochist in me as the next serious cyclist—I think you have to do what we do on a regular basis. But the joys of riding free in the wind, being one with the bike and the road, taking in nature in a 360-degree way, having your senses heightened more than they are at any other time in your daily life, sharing an experience with a riding partner, having your endorphins and adrenalin kick in, and, above all, feeling the power of subjecting the world to your will through two wheels—all this comes to mind way before the idea of punishment."

His eyes flashed as they held hers for a good few seconds. "Anything else to add to that?"

"My family raised us to indulge in good food and drink, so it helps to be able to ride that off," she said more practically. "And I will say, riding the New York streets messenger-style probably burns more calories than any other activity I've ever done."

He chuckled. "That explains why you ride as if there are no rules to the road."

"And you descend with a fearlessness I can only aspire to."

"What would you say these qualities tell us about our characters more generally?" He took a spoonful of the bouillabaisse that had just arrived.

"I'm pretty lawless. I acknowledge that as a major flaw," she confessed. "As long as it doesn't hurt anybody, I'm fine with doing it."

"And I suppose I take pride in my descending because it feels like the most technical part of what we do as cyclists," he said. "I also like throwing caution to the wind for a few moments in the day."

"Are you so cautious in the rest of your life?" she asked with some surprise.

"I think it's the curse of any disciplined life," he said ruefully. "Caution can make it hard to be present in the moment."

Nan let that statement settle for a minute as she took a bite of her steak and closed her eyes to taste its juicy rareness.

Then she said, "As far as that's concerned, I still have trouble getting the words 'aesthetic' and 'ascetic' straight and I think it's a deliberate trap on the part of our language. I mean, a sensualist and a hermit shouldn't have names that sound so similar."

Thomas regarded her appreciatively as she took another bite of steak. "Or maybe they should."

♫ ♫ ♫ ♫

They walked back under the transparent sliver of a first-quarter moon in the sky. Thomas came up to her apartment for a cup of tea, something she argued was prudent given how much they'd had to drink. "I know it's only a few miles of driving, but you never know."

Nan turned on the Bay Area's jazz station, KCSM, which she particularly loved listening to on the weekend for its original programming in classic, traditional, and Latin jazz. Even though tonight was a Wednesday, she had a feeling it would deliver.

She felt Thomas's eyes following her motions as she started the water boiling and set out their mugs. In response, she envisioned undoing the top buttons of his shirt and letting her hands rove his well-defined chest. It would also be thrilling to run her fingers along his strong jaw line and breathe in the scent of his neck. She had to steady her breaths.

They stood no more than two feet away, partly leaning against her kitchen counter, looking into each other's eyes. After what seemed like forever, Thomas closed the gap by taking a step towards Nan. In one decisive movement, he placed his arms around her waist, pulled her to him, and pressed her lips to his. Her arms naturally found their way up the sides of his solid shoulders and around his neck as she returned the firm pressure of his kiss, savoring the taste of his mouth and the smell of his skin so near. A pulsating arose between her thighs, and she gave out a small moan. Then it was over too soon, as Thomas pulled away slightly, his arms remaining wrapped around her waist.

Nan searched his eyes for answers to her questions. Did he feel they were headed for a danger zone where they would have no more control over their actions? Or did he pull away for another reason? Was it as hard for him to resist continuing to kiss her as it was hard for her to accept his decision to stop? She was suddenly too shy to consider broaching any of these intimate questions with a man she'd been daring with in almost every other way up until now. She felt she had to defer to his choice without questioning it further.

The tea kettle began to whistle, and Thomas released her from his hold. She reluctantly turned away and poured the water over the tea bags. The tension persisted, but on a lower grade.

"So, my good sir," she said teasingly, "what do you like to do in your spare time?"

"Read books in the languages I'm trying to remain fluent in and meet up with people who speak those languages in order to keep up my oral skills."

"Which languages?"

"German, first and foremost. French, Italian, and Spanish secondarily. I also try to maintain the Latin I learned in school and college."

"How in the world did you become fluent in so many languages?" asked Nan with growing reverence.

"My mom's mother is Italian and my mom insisted we all learn Italian growing up," said Thomas. "I spent a year of high school in Pescara with her relatives and became fluent there. I studied French and German intensely throughout school and, as you know, was able to perfect my German in Austria after college. I spent six months in southwestern France during my junior year at Columbia. And working amidst Spanish speakers since owning my bar has built up my confidence in my Spanish."

"And how do you otherwise find people to speak with in Italian, German, and French?"

"Mostly through Meetups where, at least before the pandemic, we gathered regularly at bars and restaurants. For the last year and a half, however, we've been meeting up mostly on Zoom."

The jazz station began to play a classic Brazilian song that Nan loved called "Tico Tico" and it reminded her of a question she'd been meaning to ask Thomas. "Do you select all the music for your bar?"

"Actually, that's my two bartenders, Elian and Luiz, who are in charge of the music," said Thomas, "which is lucky for me, because I seem to be capable of hearing only one element of music at a time when I listen to it."

"What do you mean?"

"I can focus on just the melody, for example, or just the beat, or just the emotional tone of it, but I can't seem to process them all at once. And friends tell me I miss harmonies altogether."

"So what happens when you try to sing along or dance to a piece?"

"With great concentration, I can dance, but then I'll miss out on the tune. And when I sing, people say I'm way off—but in my mind, I can hear the tune itself and it sounds good. I know I can't distinguish the layers of complex harmonies."

"But you're so good at languages, and surely that uses the same parts of the brain?"

Thomas shook his head. "There may be some overlaps, but I don't seem to have been gifted with acumen in those areas."

Nan tried not to dwell on this misfortune, the irony of which she couldn't help appreciating.

"I'm guessing you like best those kinds of music that have well-defined rhythms then?" she suggested.

"I do."

"Who are your favorite rock groups?"

"AC/DC, Aerosmith, Led Zeppelin, The Doors . . . "

"It wanted but this to confirm my suspicions that you are a man of impeccable taste, Thomas Gellert," she said appreciatively, glad to fasten once more on his strengths. "What was your college degree in?"

She handed him his tea and they proceeded to sit down on the rug in the small living room that extended off the kitchen. (Nan didn't have more than one chair and a piano bench—her apartment was truly diminutive.)

"Mechanical engineering." He sipped his tea. "I worked at a company for just three years before realizing I wanted out. It was more for my dad that I got the degree to begin with, as a kind of insurance against joblessness.”

“What did you study in Vienna?”

“I didn’t study. I worked in a restaurant and honed my German. I picked up road cycling by borrowing a bike off of one of my friends and tooling around on days when I wasn’t working. I also ended up traveling around a lot by train and bus. I decided there are really only two worthy goals in life.”

“And they are?”

Thomas smiled. “Traveling and cycling.”

Nan nodded. “Yes, I think I have to agree, those are pretty good goals. So, I’m guessing you were around 25 when you worked in Austria?”

“Yes. And it was there that I met a guy who convinced me to move out to San Francisco and work in his bar. I discovered after that that I was better at bartending and bar management than I was at food service. The gentle art of conversation that drinks encourage appealed to me.”

“And you’re happy in your career choice?”

“For the most part. Now, however, since they’ve gotten the pandemic under a little control, I’m expanding the bar into a restaurant, starting out with bar bites and gradually offering a full menu.”

“Have you got a business partner?”

Thomas smiled wryly. “No, I want to do as much of it on my own as possible. I’ve heard some horror stories about failed business partnerships. But securing a cook will be the keystone in the project.”

They sipped their teas to the sound of a trumpet-guitar-piano group playing on the radio. Nan wondered what Thomas picked out from the music, and what he felt when he heard it. Would this music feature as part of his memory of the evening, as it undoubtedly would for her?

She tentatively reached out and touched the hand that he used to prop himself up, tracing her fingers over his knuckles and the cluster of veins on the upper portion of his hand. He put down his tea and gently closed his own fingers around her wrist. He raised her hand to his mouth and placed a long, slow kiss on the inner part of the wrist where her own veins converged, holding her eyes as he did so. Then, he stood up.

“I’m glad to be my own boss and decide which nights I get to take off,” he said. “It’s been well worth it tonight.”

She walked him to the door. As they stood on the threshold, he kissed her briefly once more. "See you on Saturday," he said, turning and going down the steps.

As she listened to the stillness of the night, beyond the crickets and the background hum of cars on Highway 24, she thought she could discern the sound of a six-cylinder German engine revving up, its deep throatiness sending vibrations through the ground and announcing its raw power.

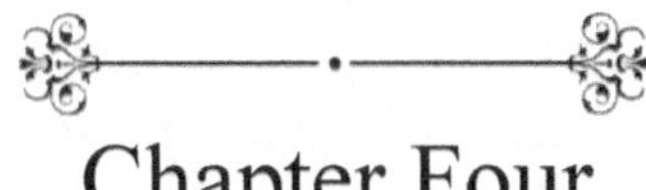

Chapter Four

As she waited outside Naseer's office on Thursday morning, Nan feared she hadn't succeeded very well at carrying out his task, although she had attempted it many times. She would observe herself seeing, hearing, smelling, and touching things in an impersonal, distanced way, without really feeling them. She would be highly aware of the time she was taking to do all of this, rather than losing herself in the present moment, as she longed to do. She would suspect she was imposing some preconceived notion on the objects she sensed. She would find herself reverting to her usual tunnel vision and her tendency to get lost in thought as she walked.

But when Naseer asked her if she'd been able to appreciate the things around her, Nan found herself saying, "I did when I was with Thomas."

"That's your cycling partner?"

"Yes. We went on a date on Wednesday. And I noticed the smells and tastes of everything. And the moon, and the music playing. And his eyelashes."

Naseer beamed, scribbling in his notepad. "So would you like to tackle something new this week?"

Nan nodded eagerly. "Yes. I want to start composing my quintet again."

"What would you say is the number one roadblock to doing that?"

"My worry that if I throw myself into one project I'll lose my skills in everything else." Nan told Naseer about her fifteen-year pursuit of excellence and her longstanding addiction to fitness. "Every week I have to do an hour of intense abdominal exercises, I have to climb at least eighty mountainous miles on the bike, I have to do two resistance workouts of between two and three hours in length, and I have to run twenty miles. I developed a system of counting all my miles exercised per day in terms of the number of calories I burn running one mile at a moderate pace. And I record every calorie I consume according to similar estimations."

Naseer playfully picked up on her own expression. "How long have you been a 'fitness addict?'"

"Since I was thirteen and I first got prescription glasses and saw my non-blurred body in the mirror. I began running four miles a day. I also tried dieting and fasting off and on through age 25. But because I've always loved food and cocktails, in college and graduate school I came to prefer exercising off the calories by taking early morning runs and bike rides after all-night parties. Then, while writing my dissertation, I discovered the gym. I found I had a natural propensity to put on muscle, and I rapidly progressed to being able to do pull-ups and dips. I loved the confidence in my own body that weight training gave me, just as I loved the way it sped up my metabolism so that it was more efficient at burning calories."

"That sounds like a secret weapon, physically and psychologically."

"Yes, only, to this day, I rarely feel I can relax into who I am as a person without thinking first about what people see when they look at my physique."

"So what if you try just focusing on composing and keeping up your fitness for awhile?" Naseer suggested. "What if you make an end date by which you'll aim to finish the composition? Something that's within reasonable scope for completing it—but that also isn't too far away?"

She reflected a moment. "We're at the beginning of November now . . . Maybe end of December?"

He nodded. "That's less than two months away. Anything you ease up on for that time period, you can always regain your strength and skills in, right?"

"I guess so."

"What if you spend the next week feeling out what it would be like to work on just composing and keeping fit—in addition to teaching? Again, don't think beyond this week for now. It's just a trial run for the two-month period you suggested."

"I think I can manage that," she said. "It's just for a week, after all."

"That's right, and then you can review the situation next week and make any necessary adjustments in other areas."

Riding home, she contemplated her goal for this week: focusing on just the composition and exercise. What if she involved other people in her project? She'd already done copious research into music composition Meetups—they didn't exist in her area—and other online groups of composers. What if she started a Meetup group in the Bay

Area for composers who were aiming to finish a concrete piece by the end of this year? This would help motivate and encourage her at this stage, and she could extend her own guidance and support to others in a similar predicament.

In reviewing the short term, Nan vowed to keep in mind the overall objective of returning to a college teaching career, where her teaching talents had shone so brilliantly, and where, she suspected, she could hold her own in composition and performance, if she only focused on those two genuine loves. *Why* had she put them off for so long? Was it because of her native perversity? She knew she had a genuine gift. It now remained only to prove it.

Over her dinner that evening, she signed up to start the new Meetup group, which she called "Bay Area Composers Composing." She proposed that their first meeting should be this upcoming Monday evening. She had one member join within the first fifteen minutes after posting.

After dinner, she pulled out the bits and pieces of her composition and began to try to make sense of them for the first time in over six months. As she sat at the piano, the phone pinged; Margot's text asked if she'd like to talk. She called Margot immediately.

"How did your date with Aaron go?" asked Nan.

"It went pretty well, I think," said Margot. "If you can believe it, that secret company he kept teasing me by text about, that he said he works for is . . . Facebook!"

"Wow. And he's a product manager, right?"

"Yes. He's not my usual type, but I feel the attraction. He's burly and nicely developed—I can tell he uses the gym and runs regularly. And then there's that sexy British accent."

"Did you guys talk a lot about disinformation and cyber-bullying?" Nan teased.

"We hit all the tough topics. He didn't directly defend the company, but he offered some interesting counter-arguments by way of hypothetically defending it."

"I can't wait to hear those!"

"But wait, before we get into that, how did *your* date go last night?"

"It was great. But I'm getting strange mixed signals from him." Nan told Margot about the two sensual moments she and Thomas had shared that had both ended in his drawing back.

"Maybe he's gotten burned in the past and wants to take this slowly so he doesn't get hurt again," suggested Margot.

"I wish he'd say something about it one way or another. It's so frustrating," Nan grumbled.

"You told me you guys had talked a lot about your past. Can't you ask him about this?"

"I guess so. It's interesting how frank he is about everything that concerns both him and me—but he doesn't get really up close and personal, if you know what I mean. Like I can't imagine him ever saying I look beautiful, or accepting a similar compliment from me about him."

"Maybe it makes him self-conscious at this stage in your relationship. Sometimes down the road, as people get to know each other, they articulate their mutual appreciation more readily. Or it could be a cultural thing."

"You are so wise, Margot. Just one last question."

"Go ahead."

"Should his taking things slowly make me worry he's not as attracted to me as I am to him?"

"I'm not even going to answer that." The smile in Margot's voice reached down the line. "I can tell—from watching you both the other night and from what you've told me—that he knows a good thing when he sees it. We have to give him some credit."

Nan laughed with relief. "You're so sweet. So you're seeing Aaron again soon?"

"Yes, on Saturday night. We're going to see the stand-up comedian Hasan Minaj at The Masonic. By the way, Aaron lives in Oakland! So I may be coming over to your neck of the woods again one of these days."

"If you ever want to have a drink before or after one of your dates, let me know. You can stop by mine anytime."

"So I'll see you at the gallery Wednesday, right?"

"What time's best?"

"Come around 3 if you can."

"I'm looking forward."

After they'd said goodnight, Nan decided to text Thomas back. She knew he was up late every night, and they'd texted back and forth a few times that day, with the general tenor of their messages light but not too flirtatious. The last one he'd sent read, "Don't go too hard on them at boot camp."

She now wrote back, "They were begging me to put them through the paces, so I did."

A minute later, she got his response: "If I were a recruit, I'd fear having you for a drill sergeant."

And they were off. She hovered over the phone for a moment, before writing, "What's your equivalent of boot camp? I know you must have one."

She didn't know whether he was going to give this a straight response or a flirty one, but either way, she was genuinely intrigued as to how Thomas managed to have the power and fitness he had when climbing. He seemed to have a secret weapon.

As if divining her thoughts, he sent back, "You want to know the real story, or the one I tell my customers?"

She texted, "Both."

"I have some HIIT workouts I follow a few times a week, plus a series of cycling-specific ab exercises and stretches. You'd recognize it all. But unlike you, I get bored with anything after an hour."

She wrote, "What do you tell the customers?"

His response: "Hauling full liquor boxes and carrying out the recycling keeps me in shape."

She had a sudden picture of him behind the bar with some good-looking female patrons hitting on him and him laughing at something they said. A stab of jealousy went through her, and she instantly tried to push it away. After all, he was texting with *her* right now at 10:30 p.m. Yet, she knew that by dint of being a bar owner, Thomas *must* flirt with women—or was that only required in the bartending phase of his career?

Her curiosity got the better of her. She texted back, "It must be the women that ask you that . . ."

There was a pause, and she feared she'd overstepped the mark. But then a text came in: "I'm much more flattered when the men ask me."

So he was an equal-opportunity flirt! But Nan remembered Thomas had said that the truth was always more important to him than what other people thought of him. Somehow that reassured the jealous part of her brain, and she resumed their thread.

"Do men hit on you a lot?" She felt more confident approaching the problem from this angle.

"Without wishing to boast, I get my fair share" came the response.

Then, after a brief pause, another text came in: "But I have the feeling that's not the real question you want to know."

Nan blushed and hovered over the screen again, not quite sure how to reply. Then she decided to take his bait. "Okay . . . do women hit on you a lot?"

"In case you hadn't noticed, I have an introverted side. And I'm always aware that the bar is my livelihood and my place of business. So I'm careful where I step," he wrote back.

She became bolder. "So they do then."

"It comes with the territory."

That way of putting it turned her on. The idea that he could show restraint in the face of constant temptation—from various sexes, no less—had her excited. In a gesture of self-defense, she reversed the current of the conversation, writing, "Would you call yourself a jealous person?"

Without hardly a pause, the text came back: "I confess that when I'm close to someone I can be very possessive."

She texted, "That's really the only kind of closeness that's worthy of the name."

"Thank you. I think so too." Then, after a beat: "So do you think it's fair to say that love dies when the ability to be jealous dies?"

Nan gave this its due consideration. "Jealousy suggests a lack of trust in the other person. Possessiveness suggests a lack of confidence in oneself to keep the other person. So I'd say that while possessiveness is healthy for love, jealousy can easily kill it."

He wrote back, "So I take it you plead not guilty to the jealous charge?"

She decided to lay her cards bare. "I'm beginning to discover that I'm most likely to be jealous at the start of something new because I don't yet know if I trust the other person."

"We'll just have to hope that the person earns your trust in time."

"Is that why the person is taking things so slowly—so both people can earn one another's trust?" she texted back before she could overthink it.

"Have you ever heard the Lou Rawls song 'Nice and Easy Does It'?" came the reply.

"Yes."

"I think those lyrics just about say it all."

♪ ♪ ♪ ♪

Nan was re-reading this text thread the next day when she got a text from Bess. It read, "Urgent. Everyone's okay, but please come over as soon as you're free."

When Nan tried to open the door to Bess's apartment, this time, for once, it was locked. She knocked more loudly than she usually did, and after a moment, the door opened to a white-faced Bess.

"What's wrong?" Nan stepped inside and hugged her sister.

Bess looked behind her and made a funny expression. "We can't talk too loudly. Come in."

All Nan saw was Bertie concentrating on a book he had open on the dining room table and Pooh-bah wagging his tail excitedly at her entrance. After greeting Bertie and Pooh-bah with kisses, Nan followed Bess into her office. Bess pulled the door nearly closed and leaned against her desk.

"I have to tell you something and you may want to take a seat," said Bess gravely.

Nan sat down on a chair by the door. "What is it?" She matched Bess's low voice.

"I didn't get sperm from a sperm bank when I was trying to conceive Bertie. I got sperm from someone I knew."

Nan was floored. "But . . . why did you tell us all you went to the Berkeley Sperm Bank?"

Bess flushed. "Because it was neater and tidier that way, and Mom and Dad wouldn't ask as many questions. And I knew that if you knew, you might leak it to them."

"Did anyone know?" said Nan, hurt as she realized who did know. "You told Teddy the truth, didn't you?"

Bess's face tinted a deeper pink in response. "He's more discrete than you are, Nan. You know that."

"Why are you telling me now?" Nan tried to tamp down her crushed feeling.

"Irina, the babysitter, was walking Bertie home from school today and she said a man followed them up to the intersection outside. She described him to me, and I knew at once who it was."

Nan's thoughts raced. "But wait, you and he must've had some formal agreement, right?"

Bess took a deep breath. "I had him sign the official form for assisted reproduction, and I had it notarized. But I've been tearing this place up looking for the form, and I can't find it anywhere. I know I put it in a really good place . . . but that was seven years ago." Bess looked desperate. As Nan surveyed the office, she realized the place was indeed a mess compared to Bess's usual tidy workspace.

"Well, surely oral agreement . . . "

Bess finished the sentence for her. "Often suffices after a thorough proof. But that costs time, energy, emotion, and money. I don't want to go that route if I don't have to."

"But are you sure that this guy—what's his name?" Nan was almost afraid to ask, now that Bertie's father had actually appeared like a ghost from beyond the grave.

"Oliver." Bess lowered her voice even further. "Oliver Morvan." She gave the last name a French pronunciation that Nan had to focus hard on ignoring for the moment.

"Are you sure that . . . Oliver . . . wants any parental rights?"

"I don't know what he wants. I don't even know what he knows. But he seems to have grasped that Bertie is his, and that we live here. He could want money, or he could want visitation rights, or something else." Bess looked questioningly at Nan.

"But Bertie isn't his!" Nan said. "Can't you take out a restraining order or something?"

"If I knew where he is, maybe. And I might be able to find out."

"Well, who is this Oliver, anyway?"

"We were good friends when I first moved out here. We met at the Episcopal Church up on Claremont—you know, St. Clement's. I used to go there a lot more before work got crazy. He attended regularly, and I found out he had a ten-year-old daughter who went to the elementary school up the street. At that time he was married. He was in the Coast Guard and had multi-year postings here on Treasure Island. But he'd been in the service for a while, and had been deployed to Alaska, Africa, the Balkans, and Central America. He was well traveled, funny, and smart. But above all, he was a really good person, I thought. He and his wife divorced around 2012, and I didn't see him at church for a while. Then when he returned a little over a year later, we started having coffees around the corner after Sunday services. He told me some amazing stories about his experiences in the Coast Guard, and said he was planning to go on one more tour in the Caribbean before retiring. That's when I was first thinking about having a child. I opened up to him about my plans for finding a donor and raising my child as a single mom."

Bess reached for a glass of water on her desk, took a few sips, and then continued. "I told him all about the IUI process that I planned to have done. We gradually came to an agreement that if I went through with this, I could count on him to provide the sperm. The following year, when my job had settled into a good groove and I'd saved up enough, I asked Oliver again if he was still willing to be a donor. He said yes. We signed the waiver, had it notarized, and then he and I got to work. He went to the clinic, ejaculated into the cup, and the laboratory washed his sperm. The physician injected the concentrated semen into my uterus, and I waited two weeks to take the blood test at the clinic that confirmed I was pregnant. You remember I had to have those polyps removed right after I conceived; but otherwise, everything went more or less smoothly, and Bertie was born the following year. Oliver had left for good during the first month of my pregnancy."

"But why did you choose Oliver, in the end?" asked Nan.

"I liked the idea of having a real, physical presence to attach to the sperm I used. After going through the profiles at the sperm bank, it all just seemed so . . . sterile and futuristic. I knew Oliver was someone who worked hard, had a strong moral sense, knew how to laugh through the toughest of times, and had a sharp mind."

"What does he look like?"

Bess got up, opened the door to check that Bertie was still absorbed in his book across the room, and returned to the desk. "Bertie has his eyes and hair . . . and chin. He's about my height, and spry. He must be 45 now."

Nan regarded her sister, who was a little over five foot ten, with blue eyes and long, thick chestnut hair. It was strange to examine her from the eyes of an unknown man, but Nan had always known her sister radiated true beauty.

"I still don't understand why you had to keep him a secret," Nan puzzled.

"For the same reasons I chose not to go with the sperm bank, people would have judged me for choosing to go with a friend."

Nan thought about that. Would she have judged her sister? She liked to think not, but she knew how much, at least in her case, the ability to withhold judgement only gradually came with age, experience, and many mistakes.

"Do you have his number? Can't you call him?" Nan suggested.

"Of course I *had* his number. But that was two phones ago." Bess held up her current Google Pixel. "I lost all my contacts that time in 2017 when I forgot the phone in the car for a whole day down in San Bernardino on one of their hottest days. It had basically melted by the time I got to it."

"You mentioned he attended St. Clement's. Maybe he'll come back there. He probably lives around here, right?"

"After he got divorced, he rented rooms in a house near the Ashby BART. But he got deployed again, so I doubt he kept those. But I am thinking of starting there."

"Do you want me to watch Bertie and Pooh-bah, or should we come with you?"

"Can you take them to Monkey Island to play? I may not be back for a few hours. I'm also going to the church."

"Okay. I'll have my phone on me."

Nan was glad she didn't have any students scheduled for the afternoon. Her fifty-something student Neal had canceled because his son was in town for the weekend, and the college-aged Charlee had begged off for today, as she was working on a biology lab report. Bertie, Pooh-bah, and Nan headed towards Claremont Boulevard, which dead-ended on a little green space the neighborhood referred to as "Monkey Island." The cul-de-sac provided a car-free haven for little kids, dogs, and adults who wanted to relax on the wooden bench beneath one of the big redwood trees. Today an elderly couple sat on the bench and a medium-sized poodle mix romped beneath the trees catching, but not retrieving, balls that his mistress threw his way. Pooh-bah and the dog greeted each other tentatively, and then Pooh-bah, who tended to be more stand-offish than most dogs, went his own way to sniff and stare down groundhog holes. Occasionally, flecks of dirt would fly up towards him, signaling that he had indeed cornered a harried groundhog. Bertie would laugh delightedly and say, "He found one! It's going this way underground, see?"

Nan observed, out of the corner of her eye, a man who strolled idly, as it seemed, along one side of the "Island." She waited until his back was towards her to survey him more directly. He certainly fit the description Bess had given of Oliver. "Bertie," Nan said quietly, "can you watch Pooh-bah for a moment while I go say hello to someone I know?"

"Okay," said Bertie.

Nan walked as casually as she could manage up to the stranger, and said, "Oliver?"

The man swung around and looked her in the eye. His look was not unfriendly. He had stubble around his jaw line and the beginnings of a mustache. His soft brown eyes and square chin did indeed resemble Bertie's, and his dark brown wavy hair was so obviously an origin for Bertie's hair that the whole effect took Nan temporarily aback.

"Yes?" he said, in a voice Nan could only describe, borrowing a term from music appreciation, as a "reedy timbre."

"I believe you know my sister, Bess. My name is Nan." She extended her hand to him.

He hesitated for a brief moment before shaking her hand. She took that as her cue to continue.

"Bess said she lost your contact information, but she would love to catch up with you, if that's possible." Nan hurried a bit as she saw, in her peripheral vision, Bertie looking towards them.

"Tell her I'll be at church on Sunday for the outdoor service at 10."

"Any other message to give her?"

Oliver gazed over towards Bertie for a long moment. "We'll see each other then. Nice to meet you." He turned on his heel and walked towards the busy Belrose Avenue.

Nan exhaled a long breath and walked back over to Bertie and Pooh-bah.

"Who was that?" Bertie asked.

"A friend of mine."

"What's his name?"

"Morvan," said Nan, not wishing to lie to Bertie.

"That's a funny name."

"It's French."

♪ ♪ ♪ ♪

"You're quiet today," said Thomas, as they climbed Spruce Avenue on their bikes towards the reservoir the next morning.

Nan thought of how Bess had called her indiscrete, and she resisted the urge to open up to Thomas about the events of the previous day. "Just some family stuff going on."

Thomas was silent for a moment. "Would you like to go to a new bar tonight that opened in downtown Berkeley? It's called The Birches. It has a large selection of local beers of all kinds."

"Beers, huh?" She smiled at him. "That's a departure from tradition for us, isn't it?"

Thomas grinned back. "In my experience, you have to take a day off from the hard stuff now and then to regain your stamina."

"Won't you be missed at The Tippler's Haunt?"

"Elian and Luiz can hold down the fort just fine. And we have a relatively new barback who needs to practice stepping up to the tougher tasks anyways."

"That sounds nice."

At 6:05, Thomas texted to say he was waiting outside her place. Nan got a decided thrill stepping into Thomas's car. Like the inside of a refrigerator, the interior of a car told one so much about a man. She ran her hand over the dark, smooth leather dashboard and the uncomplicated control panel, noting that the car was spotless and empty, utterly devoid of accessories. This was exactly what she expected from Thomas after seeing his love for speed on the bike and his simple but elegant style of dress.

"Choose whatever music you like," he said, pulling out from the curb.

For a change, Nan tuned into a classic rock station. She wanted Thomas to enjoy the music, and she wanted to drown out thoughts of Oliver. In fact, she welcomed the prospect of a noisy bar to make thinking about Oliver next to impossible.

Thomas drove exactly the way he rode the bike: with precision, clarity, and, high awareness of everything around him. Once again, she admired his parking skills as he eased into one of the few empty spots within a block of the bar.

"Some time I'd love to get driving lessons from you," Nan said, only half joking.

"You drive, right?"

"Yes, and I have an active New York driver's license; but I haven't driven seriously since graduate school. I can't vouch for my current skills."

Thomas opened the passenger door and guided her with his hand on her lower back to the entrance of the bar, pulling open the door with his other hand.

They were greeted with the sight of a couple at the bar who appeared to be devouring one another. Nan smiled quizzically at Thomas and threw her eyes skywards by way of conveying the awkwardness of the moment. A host bustled over and deliberately threw himself in the way of their line of sight, saying sheepishly, "Would you like a seat over in the corner booth here?"

After the host had seated them and stepped away, Nan and Thomas exchanged mischievous looks, letting out stifled laughs.

"I don't consider myself repressed," she remarked, "but there should be limits to PDA, right?"

"Once more, context is everything," he maintained. "That's fine for a nightclub."

"I've known people who were the extreme opposite. A guy I once dated refused to kiss, whether we were in public or in private—even when we were in bed. It was as if the mood went from zero to 150 in a second and suddenly we were having sex, without any affectionate buildup."

He shook his head. "That's like eating food without first smelling and looking at it. You miss out on most of the taste."

After establishing that Nan's knowledge and liking of beer until now consisted simply of IPAs and double IPAs, Thomas set about to broaden her horizons. Planning to steer them from light to heavy over the course of the evening, he ordered them a pilsner and a sour beer to start with. Thankfully, the bar served half-glasses, so they could pace themselves.

"Have you eaten?" asked Thomas.

"Not to speak of. I could eat."

He ordered them some sliders and crispy fried zucchini, and while they drank their water, they looked at each other, smiled, looked away, and then repeated the process.

"Do you ever feel as if you've stepped into a Richard Linklater film?" Nan said at last.

"Is that the guy who did *Before Sunrise*?" Thomas mulled this over. "I was 14 when that came out, and my mom actually let me see it, despite its R rating. I think she thought it would be good education." He chuckled.

"You *wanted* to see *Before Sunrise* at age 14?" Nan was disbelieving.

"Anything my older sisters and brother were doing was cool, and that's what they were watching."

"How did you get into the movie theater though?"

"This was a neighborhood theater where the owners knew my family. They probably even knew my parents had given the go-ahead."

"Oh. I was hoping to hear an elaborate story about how you sneaked in," she teased.

"I did sneak into the soccer stadium several times."

"Did you play?"

"Not as well as I wanted to."

"So what were you known for in high school?" She took a sip of water.

He flushed. "Liking girls."

A laugh rippled out of her. "Okay, now I'm beginning to understand the *Before Sunrise* bit."

They received their beers and tilted them up to the light to admire their complexion.

"I don't really know what I'm looking for here," she admitted, "but I've seen people do this."

"Salud," he said. "I'm interested to hear your thoughts."

Nan tasted both, liking the grainy aftertaste of the pilsner more than the tartness of the sour ale.

"I'm guessing you were also a science and math geek in high school," she said. "Am I right?"

He nodded. "And chess, and debate."

"Wow, two things I steered clear of. I was a quiz bowl and theater person."

"That I can readily believe."

Their food arrived, and they each took a piece of fried zucchini, dipping it in the accompanying Green Goddess sauce.

"Were you into sci-fi?" she asked.

"I preferred literary fiction to sci-fi and fantasy, even then," he said. "And to this day, if I have a choice between reading the philosophy of an author and his or her biography, I still choose the philosophy."

"That *I* can readily believe. You have a very intense side, Mr. Tippler's Haunt."

"You're suggesting my laid-back side is the default."

"You yourself referred the other night to your withdrawn side."

"There are times when I really don't want to be around people."

"What do you do at those times?"

"Exercise, read, practice languages, and cook."

"Now you're just trying to make me swoon. You cook on top of everything else?"

"If by 'everything else' you mean mixing a good drink or directing people to a good glass of wine or beer, then yes."

She tilted her head at him meaningfully. "You know by 'everything else' I mean a lot more than that. First off, you can do things on the bike that, as far as I can tell, only serious racers can do—like putting on a jacket while riding with no hands. Second, you fix everything on your bike yourself. I take mine into the shop as soon as something goes wrong—all I can do is fix a flat, use the barrel adjusters to do basic adjustments, and tighten up the odd bolt." Nan took a sip of sour beer and continued, warming to her subject. "Third, you're obviously a lot more conversant with poetry and philosophy than you let on. Fourth, you know at least five foreign languages. I could go on, but I fear I may make your head swell up, as my grandmother used to say." She paused, and then added more quietly, looking down at her beer, "But I will say you know how to kiss."

"That may well be worth more to me than all the rest of the lot." He laughed.

They made their way gradually through a pale ale and a saison to a porter and a stout, and Nan clutched her tummy. "You should've warned me: beers are like an entire extra meal on top of the one you're eating."

"People in Germany and Austria have beer not with, but for, breakfast. It seems like an entirely healthful way to begin the day."

Even as she had the thought, Nan knew it was clichéd, but she realized she dreaded having the night end. The gentle buzz of beer was entirely different from the stronger kick of cocktails, and she appreciated the chance to absorb more of Thomas from a calmer, less frenetic high. She saw him lower his eyes to her hands several times in the course of the evening and at last she asked him, "What is it?"

"Your fingers look like a pianist's fingers," he said simply.

"What do a pianist's fingers look like?" she prompted, amused.

"Long, slender, graceful, capable . . . "

Nan looked down at her fingers and, as she did so, Thomas traced his own index finger over each of hers lingeringly. As he'd done at her apartment, he met and held her eyes as he performed this movement, and it was the keen light that shone from his own eyes that allowed her to see into his mind. She saw, in that moment, that he did indeed want her very much. The conviction of his strength of feeling made her suddenly shy and quiet.

A memory of what she and Margot had talked about the other night recurred to her. "Do you think dating in multi-cultural settings is particularly challenging?"

"How so?"

"Navigating your way through utterly different cultures and the ways they have of courting, I mean," she clarified. "And not just the ways they court, but the ways they experience personal space, the ways they see time, the ways they express intimacy . . . "

He nodded. "Two people from the same cultural background must have an advantage—at least, at first."

"And then later?"

"Over time—if a bi-cultural couple make it past courting—I think their differences make them stronger, individually and together. If they're able to work through the different languages involved in being intimate with one another, arguing with one another, teasing one another, they both become effectively bilingual."

The notion impressed her. "I guess anyone who's the product of those couples has their parents' bilingualism woven into their fabric from early on. But do you think most people still seek out partners from their same culture and resist facing the bigger challenges of dating a person from another culture?"

"Without a doubt—or, if they do venture out of their culture, if they don't then put in the necessary work to learn and appreciate their partner's differences, they split up and mistake it as a lesson that they ought to have stayed within their own culture. But bear in mind, I speak merely from observation and anecdotal evidence. He finished his beer, smiling impudently at her. "Would the encounter of Hungarian-American and Scottish-American cultures figure as Exhibit A in this study?"

She shook her head. "No, I think you and I have lived through and seen enough in our time . . . this is probably more like Exhibit O."

He roared with laughter. "Well, I look forward to seeing the results of Exhibit O in time. Shall we square the bill and then perhaps add some touches to our study?"

"Now you sound French," she said, but then wished she hadn't, since it brought up thoughts of Oliver again.

"A dubious compliment at best, but we Hungarians bear our French brothers no ill will."

When they stepped outside, the cold November air felt like a punch to the system. Nan shivered and leaned into Thomas, who put a warm arm around her. "Would you like my scarf?" he asked, and without waiting for her reply, he wrapped it around her neck.

"Ohhh, that feels so good," she sighed. "But you'll catch cold."

"I always bring it just as insurance—I rarely actually need it."

They walked slowly to the car—was it Nan's imagination, or were they walking much more slowly than they usually did? Cold as it was, she, for one, wanted to linger out the short walk as long as possible.

When they reached the passenger's side of the car, Thomas turned to her and cupped her face in his hands. He let the suspense build for a moment—endless moment!—and then met her waiting lips with his. He parted the way for his tongue, and, with the briefest of touches, made contact with her tongue. She was preparing to respond more fully, when he drew away slightly and stood, his face not two inches from hers, his eyes searching her own. She wondered if her eyes told too plainly her thoughts: why such restraint? How can you pull away now?

"If you're trying to drive me crazy, it's working," she grumbled.

He tucked a strand of hair behind her ear. "No 'sweet reluctant amorous delay for you,' I know," he said with a half-smile. Then he leaned in to whisper in her ear, "But that was never attractive anyways."

She shivered from the touch of his breath in one of her most sensitive spots. "You *are* trying to drive me crazy!" she said with sudden realization.

He looked amused. "You said yourself the other night, we're both masochists. Let's get you warm." He stepped back and opened the car door for her.

Nan grudgingly stepped in, giving him a baleful look as she did so. "In case you didn't notice, you just gave away your game."

"In this one area I avoid games, actually," he said. "I'm a straight shooter."

"Hmmph," she muttered. She waited until he'd gotten into the driver's seat, and said, "You want to be honest about what's on your mind then, straight shooter?"

He turned the key in the ignition and the engine roared to life. "When I say I enjoy the long seduction, I mean I think of it as an art in itself. An often neglected art."

"Art—game, what's the difference here?" she challenged.

He turned to her. "Games are directed purely towards enjoyment and they usually have winners. Art aims foremost at beauty—even if ugly beauty—and its pleasures are longer-lasting."

"How can you confidently guarantee your 'long seduction' will have no winners—or losers?"

"You, of all people, wouldn't waste your time on any seduction that didn't have dangers," he said, meeting her eyes frankly. "You thrive on danger."

"But who's the artist behind all of this seduction? Who gets to call the shots?"

"We both do," he said. "With good communication."

"Speaking of good communication, I'm glad you finally decided to let me in on this work of art we're both making."

He seemed startled by her sarcastic tone. "Nan, we're still earning each other's trust, and trust isn't something that comes easily for people who are as much in their heads as you and I are. But we have going for us the fact that we also feel deeply and have a strong dose of impulsiveness that lets us leap into things that seem worthwhile. Doesn't intuition tell you that I wouldn't intentionally hurt you?"

"What you said before makes more sense to me. Direct communication is going to earn my trust." She hesitated a moment before adding, "But can we talk directly about this danger business? What kinds of dangers did you have in mind?"

Thomas's eyes flashed. "I knew I'd read you correctly."

♪ ♪ ♪ ♪

With the greatest difficulty Nan taught her 11 a.m. class the next morning. So much seemed to be happening all at once that it was hard to concentrate, even as her fifteen-year-old student Chloe played through the first movement of one of Mozart's most beautiful piano sonatas, the sonata in F. She knew that at this very moment, Bess and Oliver must be convening at Peet's Coffee on Domingo near the church. Irina was watching Bertie for the duration, but the outcome kept Nan on edge. How long would it be before her sister texted her?

She also had her first Meetup with fellow composers scheduled for the following evening, and she tried in vain to keep from planning out what the shape of it should be while giving Chloe her lesson. She'd gotten eight definite attendees, and five more members who couldn't make the first meeting—all in all, a number she considered a good start to their group. She wanted the structure of their meetings to be somewhat free-form, but for them to have a definite timeline in common for reaching their goals.

Once she and Chloe had established a few goals for Chloe to tackle in her practice during the following week, Nan saw her to the door, and realized she had fifteen precious minutes in which to make a cup of tea before her next student arrived. She texted Bess to ask if everything was okay.

Now, at last, the conversation with Thomas the night before replayed in her head. As he'd driven her home, it had transpired that Thomas had picked up on many small cues over the course of the past year while riding with Nan, and he had retained them all without having misinterpreted a single one. Nan was still reeling from some of the observations he'd shared about her riding style, her comments during their exchanges, her body language on and off the bike, and her cultural attitude towards intimacy, much of which he'd gleaned from the smallest details of her behavior. The upshot of his penetrating reflections was that Nan was that rare person of power and strength who sought continually to contend with another person of equal or greater power and strength in order to become better. He accurately discerned that while this competitive instinct towards improvement guided her daily behavior, where lovemaking was concerned, she had slightly different expectations of her rival. There she wanted to be guided as in a dance and to yield to the seductive powers of a man who represented just enough of the unknown to be somewhat dangerous. This rival must be prepared to take control whenever she willingly ceded it, including in those social situations where he naturally surpassed her in knowledge or ability, or where convention cast him as leader.

Thomas was careful to distinguish the potent desires he perceived in her from role-playing, which he claimed was a greatly watered-down concept that didn't depend enough on individual character. To him, the standard simplistic categories of dominance and submissiveness, for instance, were at once too violent and not forceful enough for the intellectual eroticism that they had established early on in their flirtation. Nan's imagination, like his, was capacious enough that it could remake reality, with the right person. This process of recasting reality made up the "art" of the long seduction, in which both artists got to know their subject well enough that they became their subject.

Nan had been turning over this last bit since Thomas had dropped her at her place and kissed her goodnight. She wondered what it really meant to "become your subject" when you were creating a work of art. Was it a sort of Pygmalion moment in which your subject came alive for you? Presumably, if lovemaking was to be conceived of as an art, it must be a constantly changing work of art—otherwise no man could continue to represent the unknown long enough to remain dangerous, and no woman could fascinate long enough to make it worth his while. But, she supposed, when imagination worked in conjunction with desire, it was entirely possible to transform reality. Nan also wondered about the nature of Thomas's desires. Surprisingly, she was not concerned with a question she felt might be uppermost on the minds of observers—that is, whether Thomas had always only wanted a sexual relationship with her. Nor did it seem presumptuous in him to have independently extrapolated as much as he had about her. Somehow, when he shared these thoughts with her, he spoke divinity that only to others might seem profanation.

A frightening thought struck Nan: was she falling in love with Thomas? How could this be possible so soon? They had barely kissed three times. She re-examined the highlights of the past week or so, since the moment she'd thought of visiting his bar. These high points combined with flashbacks of Thomas on the bike, conversations they'd had on the phone over the last year, and intimate moments they'd had when they'd stopped while riding. Shouldn't the clock on their relationship be set back almost a year? It now appeared they knew a great deal more about each other than she'd initially supposed, and this knowledge enabled her to trust him far more than a week's time would warrant.

She surveyed her own state of mind. It had not escaped her notice that for the past week a lightness had returned to many of the things she did. She had been more full of good humor, patience with annoying people, and forgiveness for her own vulnerabilities as well as those of others. She'd surprised herself with her readiness to accept Bess's charge of indiscretion; her tolerance when Thomas had revealed his limitations where music, of all things, was concerned; and her ability to overcome her self-doubt when her dad had announced Teddy's engagement. For four nights in a row she had even relaxed her bedtime without obsessing over the potential outcome. She was finally taking the first small steps towards writing her quintet, using therapy to remove some key mental and behavioral barriers to happiness, and plotting how to scale back slightly on her relentless pursuit of excellence.

If she was falling in love, what then? She would have to dissemble with Thomas, for it was inconceivable that he could have reached the same point so early. Perhaps by feigning less intensity than she actually felt, she could even bring herself to a cooler position, one both more convenient and more acceptable for this stage in their friendship.

The surest sign that Nan was smitten was that she itched to write poetry. She couldn't wait until all of her students had had their lessons, and she'd heard word that Bess and Bertie were safe, to jot down some of the phrases that had been blazing like meteors through her head since the previous night.

Once she'd taught Harley in the mid-afternoon, and she had a half-hour lull, the poetic urge overcame her, and she took a moment to write out the following sonnet:

Let me tell you about desire, you of equality and moderation.
I know all about desire for love, for friends, for music,
For critical acclaim, for times of relaxation,
For Portugal, Morocco, New Zealand, Thailand, Dublin,
For movies, exhaustion, the twisted mind that's worked itself to
    death,
And succeeded at what it's spent itself on,
And looks longingly back leagues into the past.
I can tell you of frustration and satisfaction
When lovemaking, drinking, running, writing, conversing;
I can communicate emotions of chocolate,
The urge to squeeze, wrench, tug, destroy, deflower, unthread,
And restless, relentless recountings to one or another
Who know nothing of striving or straining, who perhaps cannot
    admire,
But who can only stand to take in, at a distance, my overwhelming
    desire.

She supposed the natural title to the poem was "Desire." As she was finishing scribbling this across the top of the sheet of paper, a text came in from Bess:

"Finish teaching your classes. All is okay, but there are complications."

Nan texted, "I'll be over at 7."

When Nan had finished giving her last lesson at 6:45, she first shot Margot a text: "Hope to catch up with you later about your date last night!"

Margot replied, "Back atcha!"

Nan rode her road bike to Bess's place, knocked, and tried the door. This time it was open, which Nan took as a good sign.

"Bertie?" Nan looked around and at last saw that Bertie was cuddled up with Pooh-bah on the sofa playing a game on his iPad. Pooh-bah's tail thumped enthusiastically. Nan engulfed each of them in a hug, before asking, "Where's your mom?"

Bertie said, "I think she's in the bathroom."

Bess came out a second later. "Nan, would you like a cup of tea?"

She made them both a cup, and they retreated once more to Bess's office.

"Oliver hasn't changed much," Bess observed by way of an opener. "He's still sensitive, funny, and warm. He complimented me on Bertie right away and said he seemed well raised. He retired from the Coast Guard in August and took the last few months to visit family in Detroit. Now he's decided to settle back here in Berkeley for good. He said that while he was in the Caribbean he started having 'spiritual wonderings.' I asked him what he meant, and he said that when you're out on a boat for a long time, you start questioning the meaning of life, God, existence, and so forth. He saw a vicar at a church in the Virgin Islands, a Reverend Charles Applethorpe, who said that he should take responsibility for his family. He told this Rev. Applethorpe that he thought he had a son whom he hadn't yet met, and apparently this sparked the vicar to advise him to come meet his son and provide any support and guidance he could by way of a father figure. This was a year ago, and now Oliver has come to find Bertie. I'm getting the sense that Oliver is lonely, now that his daughter Angela is in college and a lot of his friends have moved away. He's casting about for any friendships he can reignite from the past."

Bess took a few sips of tea, and then continued. "Oliver came out and asked me if Bertie was his."

Nan held her breath. "What did you say?"

"I said no. I said that the IUI had resulted in a miscarriage, and then after Oliver had left for his tour, I'd succeeded in getting pregnant through a donor from the sperm bank."

Nan considered this. "But it's striking that Oliver doesn't seem to have any mercenary motives. I mean, he seems genuinely to want to be there in case you or Bertie needs him."

Bess flared up at her. "We don't need anything from him! We don't need him in our lives at all."

Nan reasoned, "Okay, but he's going to be around a lot now, and he's liable to realize you've told him a lie, and Bertie is likely to ask about his father soon . . ."

"I can defer that a lot longer. I don't need to do this on anyone's timeline but my own."

"But he seems like a truly good person, and he could be nice to have around . . ."

Once again, Bess was seething. "Nan, Bertie and I are fine without anybody else. And if you don't think so, maybe you should stay away too. We don't even need you."

This stung Nan into silence.

"Did Oliver say something to brainwash you?" Bess persisted. "Because it sounds as if he's hoodwinked you into taking his side."

"No, but he seems honest enough. Don't you think you should be honest with him? Not for his sake, but for yours."

"Nan, you don't know what it is to have a child. You may never know. You're so wrapped up in all your selfish artistic pursuits, you never have time for human relationships."

Once again, Nan was cut to the quick. She got up.

"I'm leaving," she said, a lump forming in her throat. "You know how I feel about this. Maybe we can talk about it some other time, when you're not so angry."

"Go ahead and go," said Bess. "We won't miss you."

Chapter Five

Margot and Nan had a two-hour talk that night when Nan returned from Bess's. Margot told her all about her date with Aaron, the stand-up show they'd seen, and the hours they'd spent drinking tea at her place afterwards. They'd kissed a lot, according to Margot, and she appreciated the depth and well-roundedness of his conversation.

"I think it's really true what they say about British high school—it is like our college, and their college is equal to our graduate school. It's so unfair," said Margot.

Nan, in turn, told her all about Saturday night with Thomas, his concept of the long seduction as art, and her burgeoning feelings for him.

"It's such a complex idea—I don't fully understand it myself yet," Nan admitted. "On the one hand, it seems that the 'long seduction' assumes that contest and struggle form the pleasure of love. Part of that struggle includes holding back on lovemaking. On the other hand, the two minds that create this 'long seduction' are always doubting themselves and each other. And in the midst of this, Thomas and I have a particular dynamic that demands that the winner of the conflict must always be the one who shows the greater exertion of will. Since I love to be subdued by him, it's not hard to see that the conquest is likely to go to him."

She didn't know how Margot would react to all this, but she had determined not to omit a single part. She had no cause for fear, though, since Margot, perspicacious as she was, picked up immediately on the uniqueness of the situation.

"What if you two were absolutely right for each other? What are the odds that you'd ever meet and communicate your singular inclinations, or that you'd both be the sort of people who are willing to act on them?"

Nan loved the way Margot never worried about what other people thought, the way she discarded the conventions that didn't suit her needs, and the way, rational as she was, she believed in following one's heart. She also tended to give people the benefit of the doubt, an all-too-rare quality, as Nan was discovering.

Nan told her about her falling-out with Bess over Oliver. Margot consoled her and reminded her that all sisters have their share of disagreements, but that she was sure Bess would come to understand her point of view in time. "We can't know what it's like to be in your sister's skin," said Margot. "Just be there for her when she needs you again, and your differences will sort themselves out."

Monday evening, Nan welcomed the chance to hold her Bay Area Composers Composing Meetup. She planned to take fifteen minutes at the beginning of each meeting to talk about compositional challenges members might have in common, and then for them to work silently for an hour on their individual works. During the last fifteen minutes, they had an opportunity to share what they'd been working on. Some people, she knew, preferred to compose with pen and paper, while others composed on the computer. Some people had programs that played the composition directly from their computer; others could share via instrument or vocals. Anyone who wanted to could share their screen with the group.

The group had a range of ages and levels of ambition. A handful were young composers in the Los Angeles area who'd jumped at the chance of a California-based group supporting composition generally. Apparently, most composition groups in L.A. were focused on movie music, or at least on tailoring one's compositions to the film industry. There were a couple of older folks who were following earlier dreams of finishing compositions left unexplored for years. Then there were two Bay Area composers who were music students wanting, like Nan, to beef up their CVs.

Nan proposed they aim for December 31 as a deadline for producing something complete that might be shared with the other group members, and she took a vote. Everyone except Todd, an elderly man who couldn't seem to work the Zoom vote function, weighed in favorably. Nan suggested that after this first week, they could begin each meeting by discussing an element of compositional theory or craft. She then offered a few musical stimuli which she'd used with her students at Harvard and Drummond to inspire creativity. She proposed varying the stimuli over the first few weeks depending on what they found useful—whether rhythmic, melodic, harmonic, or extra-musical stimuli.

Because of introductions and the laying out of the Meetup's projected goals, they didn't get started writing independently until about half an hour into the ninety-minute meeting. Nan gave them 45 minutes to work, and then during the last fifteen minutes she re-convened them for optional sharing out. A young woman named Paige, who was based out of a suburb of L.A., shared an inventive folk piece that lasted a mere minute, but one that Nan was excited to see developed further. Bruce, a thirty-something music graduate student at UC San Francisco, played on his piano various chord progressions of a modern classical composition that he planned to flesh out into an orchestral piece. After giving everyone else a chance to share, Nan played some of the parts she'd built up for piano, two violins, viola, and cello. While her style was modern, it brimmed with echoes of some of her favorite composers from the last century: Debussy, Ravel, Florence Price, Gershwin. She was composing the first movement in her preferred key—C-sharp minor. She thought the second movement might be in E-major, and then the final movement would alternate, as Beethoven's "Ghost" trio did, between the two related keys.

By the time the group had said their goodbyes and signed off the Zoom meeting, Nan was exhausted. She'd taken a trail run before the two piano lessons that she'd held before the Meetup, but hadn't had time to eat since lunch. Now it was after seven and she was torn between needing to eat and just wanting to fall into bed. As she opened the refrigerator to survey its meager contents, her phone pinged. It was a text from Thomas.

"How did your Meetup go?"

"Great!! I'm starving now though."

"Let me come pick you up. I've made risotto, and it needs sharing."

"Do you need any other ingredients to finish it?"

"Just you."

Nan smiled. "Okay."

Nan gambled that Thomas's place would be warm enough that she could get away with wearing just the thin cotton leggings she had on now and one of the Merino wool base layers she wore close to her skin at all times in the Bay Area evenings. She rooted around in her liquor cabinet below the counter and found a bottle of Chablis that she figured they could chill in Thomas's freezer for half an hour.

As she waited outside for him to arrive, Nan was glad she had no time to overthink things. She had a restless energy and a heightened sense of things around her—the waxing moon that hung low in the sky, the mutterings of a homeless man who pushed his cart on the opposite side of College Avenue, the brilliant-gold ginkgo leaves that had reached their peak on the tree by the bus stop, the sharp bite of the moist air, the savory smell of grilled steak that wafted over from a restaurant across the street. This last made Nan's stomach rumble, and by way of distraction from her hunger she sang a snatch of melody from her projected third movement. She didn't know where she could take this theme after about sixteen notes, but she planned to set aside at least two hours at the start of each day to work on the composition.

Thomas's car pulled up and she smiled as she opened the door to climb in. Thankfully, he had the heat on. He leaned over and kissed her on the cheek. "Nice to see you."

"You too," she said, breathing in his smell. "What's going on at the bar?"

"Mondays are always slow. I usually take them off."

"That's lucky for me. You called at just the right time."

She reminded herself, as they drove, to keep her cards close to her chest. She may have been giddy with excitement about seeing Thomas's house, but she planned to play it cool.

"What's that tune you're singing?"

She looked questioningly at him before realizing she must've been humming out loud. "It's part of the piece I'm composing."

"It sounds good."

Thomas's house was spacious, light, and airy. The lofty living room led directly into a dining room which, in turn, flowed into an open kitchen edged on one side by a built-in bar counter. Nan noticed a wood-burning fireplace on the wall to the left. At Thomas's wave of invitation, Nan explored the two bedrooms and the bathroom, noting immediately that the latter had a lovely claw-foot bathtub. (Her own bathroom was the size of a cruise ship bathroom, offering only a small plastic box for a shower.) Thomas used the front bedroom for an office, in which he kept a clean wooden standing desk, a laptop, two chairs, a bookshelf, and some exercise equipment. Nan noted the pull-up bars in the closet doorway and the rowing machine in the corner. She cautiously opened the door to the back bedroom. It had a skylight in the middle of the high ceiling and a huge king-sized bed covered neatly with a dark duvet. Trying to look without actually looking, Nan quickly took in the bookshelf, dresser, and closet. Everything was neat, dustless, and sparse without being hermitic or cold. She pushed the door to and joined Thomas in the kitchen.

After placing her white wine in his freezer, she said, "What can I do to help?"

She was facing the counter and Thomas reached around her to open up the cupboard. As he did so, he placed a hand on her lower back. He pulled out two pasta dishes and laid them on the counter, remaining just behind her for one exhilarating moment longer than expected. As Nan imagined him pressing his body against hers, she tried to keep a shudder of pleasure from giving her away, and she took a step to the side to control herself.

He offered, "If you'd like to put on some music, my iPod is over there on the dining room table. It's hooked up via patch cord to speakers. But if you'd rather use your phone, they are smart speakers."

She scrolled through his playlists, selecting one that said "Latin." It seemed to have a mix of Cuban, Puerto Rican, Colombian, Venezuelan, and Brazilian music.

The complex rhythms of a Venezuelan song blared out, and she turned it down slightly.

"Come, sit." On the bar counter Thomas had set out two dishes of piping-hot risotto and two wine glasses next to them. "We'll open your wine after we've finished this bottle I used for cooking the rice. It's a Carricante from Sicily."

They took their seats on the two high bar stools that Thomas had placed next to one another, and Nan inhaled reverently. "Wow, that smells heavenly."

"Don't stand on ceremony. As the Italians say, 'Mangia, mangia.'"

He watched her as she took a bite and closed her eyes to savor the subtle flavors of the porcini mushrooms and thyme. "Ohhh," she groaned. "This is so good."

She took a sip of the wine, which was medium dry and acidic, a perfect complement to the risotto and the Parmeggiano-Reggiano Thomas had grated on top of it.

"Have you ever thought of being your own chef at your bar? This kind of meal would draw food-lovers from all around the Bay Area. Only," she reflected, "then you might have to change your name from The Tippler's Haunt to The Gourmand's Corner or something like that."

He smiled. "I love cooking for friends and family. But I think it would spoil it to have to cook for strangers."

"Your house is incredible," she said, finishing her first helping of the risotto and eyeing with longing the pan where more remained. "I see you have a washer and dryer—and that, like me, you don't own a T.V."

"I never have. Some time in my late teens I began to notice the only good things on T.V. were reruns of old shows. Now, like millions of other people, I simply stream things on my computer. I do have a 24-inch monitor I can hook up to the laptop in the living room. I do this, for example, when one of the grand tours is on."

"Do you have Netflix?"

He shook his head. "I have ways of streaming things for free without getting viruses or malware on my computer."

"My God, are you also a tech genius?" Her eyes widened.

"Far from it. But I know how to do what I need to when it comes to computers and electronics." He took her plate and said, "More?" She nodded.

Thomas removed her wine from the freezer, opened it, and poured them both full glasses. He refilled their dishes with more rice and handed them over the counter.

"What movie would you see next—that's streamable?" she asked.

"*No Time to Die.*" At her blank look, he added, "The latest James Bond flick."

"Oh, I like Daniel Craig in the role of Bond. I haven't seen one of those films in too long. Come to think of it, I haven't seen any film in way too long."

He raised his eyebrows. "Would you like to watch it after dinner?"

"I might fall asleep while we watch . . ."

"If you are able to fall asleep during this film, I'll credit you with being a true narcoleptic."

After they'd cleared away the dishes, they took their wine glasses and the rest of the bottle over to the sofa, and she helped him clear space on the coffee table for the monitor, which he brought out and hooked up to his laptop in under ten minutes. He called up the movie online, and took a seat next to her. She was aware of the dense sinews of his thighs beneath his sweatpants and couldn't help admiring his shapely narrow feet, which were bare on the hardwood floor.

"You're not cold?" she asked, looking up again.

"Far from it," he said. "Are you? I have a blanket here."

"Sure, I'll take it."

As he reached around for the blanket, she caught a glimpse, through his white tee-shirt, of his muscular abdomen and the striations of his obliques. She sighed involuntarily. This was a dangerous situation she'd gotten herself into. How long could she keep herself from touching Thomas or from giving away her intense longing for him?

As he started the film, she said in a low voice, "This is going to be pure torture."

He put an arm around her shoulders and traced his fingers lightly over her collarbone. "It doesn't have to be *pure* torture. We can make it whatever kind of torture we want."

"But it does have to be torture, I suppose?" She turned towards him.

His lips hovered an inch away from hers. Matching her low tone, he said, "It's always only been about what you and I want. That's what makes it exciting."

She studied his full lips, and then moved up, scanning his strongly defined nose, his deep hazel eyes, his dark brown eyebrows, his broad forehead, his thick black hair. "I think you should pause the movie. I don't want us to miss anything."

He tapped the trackpad on the laptop without moving his arm from around her shoulder.

"Remember how you said that competition drives what I do every day, but that when it comes to romantic relationships I crave a power differential that favors my partner?"

He held her gaze by way of affirmation.

She hesitated, taking a deep breath. "I'm finding myself struggling like a bird in lime because I think I may be falling to my captor. And my survival instinct is kicking in." Her voice trembled and she held her chin high as she said this, meeting his eyes bravely.

Thomas regarded her with the utmost softness. "If you're afraid your captor would do anything to harm you, you forget how captors can themselves be captive to their prey."

She tentatively ran her fingers over the top of his tee-shirt where his pectorals began. "I'm not so much afraid of his harming me as of his not knowing his own heart as well as I know mine."

"What if your captor should confess that he's been falling for his prey for some time now?"

"But wouldn't that confession remove the danger of their encounters?"

"He's still the captor," he said, eyes glinting. "And his prey is still the prey."

♪ ♪ ♪ ♪

They made it through the movie without Nan falling asleep, although leaning her head on Thomas's chest made her infinitely more relaxed than she would have been in her own bed. It was engrossing action, and she worriedly posed the question at the end, "Is this really the end of Bond?"

He shook his head. "They always leave an out for James Bond. He'll be back, against all the odds."

Tired as she was, she wanted them to remain as they were for a good long while. Despite the course of their earlier conversation, she felt safe next to Thomas and realized she valued this security as much as the danger they'd discussed so much of late. "Can I ask you something?" she murmured sleepily.

He nodded, his hand stroking her hair.

"Do you think we're a couple of nutcases?"

He laughed, the deep rumble in his chest soothing her spirit. "You mean because we don't do things as other people do?"

"Here we are, two people who are intensely attracted to each other, with no social or moral restrictions on our behavior, and we create our own obstacles for being together. Isn't that the definition of a screwball?"

"Delayed gratification isn't so insane."

"But can you explain to me again why we do it?" she persisted.

"You mean beyond the fact that it's hot?" He smiled. "Deferral sharpens our desire for each other. It builds trust, and ensures we're emotionally in sync as we develop our relationship. Not least of all, it heightens the danger of our encounters, since it allows our imagination to conjure a reality based more and more on what we know of each other."

"But how do we know that it will all be worth it in the end? What if we fail miserably and the whole thing blows up in our faces?"

"The beauty of this sort of alliance is that it can't fail, not so long as we invest wholly in it—and in each other."

Nan thought about that for a moment. "So you're saying that by giving ourselves over completely to the vision of our long seduction, our bond becomes unbreakable—and that only reinforces the vision?"

"I prefer 'underlying idea' to 'vision,' but yes, that's the essence of becoming the subject of your art, in this context."

"Can we use words to complement actions in this long seduction?"

"We already have, on many occasions—on and off the bike."

"I mean, could we take turns building an image with words that the other person responds to through a physical action?"

Thomas moved his free hand to the front of her neck, cradling it suggestively between his thumb and forefingers. She quivered from the destructive potential behind the gesture and its tantalizing intimacy.

"Or," she said shakily, "we could build off of a physical action to make our verbal image . . ."

Thomas nodded. "You go first."

Nan began:

"How I love

Your denial of imperfection to the end

When an instant poised between the cut and blow

Determines what you shall perform for me

And for yourself."

Thomas ran his fingers gently along the inner thigh of the leg she had crossed uppermost on the couch. She uncrossed her leg and moaned at the same time, leaning more deeply into him as she did so. She recovered enough to object, "You don't play fair! You went out of turn."

"You created an arresting image," he replied equably. "I responded in kind."

"It's time for you to speak," she challenged.

Thomas reflected for a few minutes. Then:

"That was the sound of hunger.

In our wheels and their spin

Circuit, cycle, span of life

Ecstasy, terror, lull, and din."

Her eyes flashed. Then she did what she'd been wanting to do for some time now. She turned his face to hers, tilted her mouth towards him, and pulled him into a long, warm kiss. With eyes closed, she invited his tongue to brush and tangle with hers, and his tongue responded playfully and passionately. She saw nothing but deep purple as they both sustained this moment before pulling slowly away.

She couldn't help but notice the rigidity that had risen between his legs. She herself was oozing with desire and a steady throbbing thrummed between her thighs. So much for playing her cards close to her chest, she thought briefly.

She retreated to the space between his shoulder and chest, and nestled her head into its deep solidity. "Personally, I think that's all the danger I can handle for tonight," she said, registering the sped-up heartbeats in his chest.

They sat like that for a few minutes in silence, Nan thrilled by how close to the precipice they'd both come and not daring to disturb Thomas's thoughts with any more words. Presently, his heart rate slowed to a steady pulse, and she herself felt more composed.

"I'll drive you home," he said, removing his arm from around her.

They remained quiet while he drove them through the empty streets. As they pulled up in front of her place, she broke the silence. "Thank you for dinner and . . . everything."

He smiled. "My pleasure. See you Wednesday?"

"See you Wednesday."

♪ ♪ ♪ ♪

Nan had two consecutive dreams that night, first of a scene in which Thomas had backed her up against the wall, his arms encircling her head and his pupils dilated with desire, and then of a scene in which she overheard Bess and Oliver talking at a table in a coffee shop about how

Bess was planning to take Bertie on a long trip out of the country. She awoke from both of these perspiring and with an accelerated heart rate.

She lay awake for more than two hours before getting back to sleep. In this time, she seemed to compose an infinite number of variations on her initial theme for the third movement of the quintet, and none of them were adequate. When she awoke Tuesday morning she felt as if she'd hardly slept at all. Despite her sluggishness, after making a four-shot espresso with her machine, she felt relief resorting to the computer to tackle two hours of composition. She jotted down notes on a staffed notepad of the six major motifs for her piece and of the places where these might fit in.

After writing for two hours, Nan decided to change up her predictable breakfast routine of soft-boiled eggs and toast. She made cornmeal pancakes from *Joy of Cooking* and, in the course of frying up two of them side by side, noticed what looked like an orca cavorting across one of them. She laughed to think how this creature had formed from the mixture and a series of random acts on her part as she'd poured the mixture. Was this all humans were, after all? The casual result of an unplanned series of moves on the part of a higher being?

Nan had called Bess twice since Sunday, but the phone had gone over to voicemail. She didn't really know what she would have said had Bess picked up, for she hadn't changed her position concerning Oliver. But she wanted to leave the channels open for her sister to talk. It didn't seem real yet that they had had a falling-out. Nan kept considering the problem from every angle. How had Bess managed to discount for all this time the very real possibility that Oliver would one day return to Berkeley and that they'd find themselves in this predicament? Would she be more inclined to be charitable towards Oliver if she were able to find the form that Oliver and she had signed seven years ago? Would Bertie benefit from knowing who his father was?

After doing a two-hour weight workout in the back driveway using kettlebells, dumbbells, and her own bodyweight, Nan spent the rest of the midday advertising for more students on various sites. She held a lovely piano lesson with a twelve-year-old named Joanna, who was working on the first and third pieces in Fauré's "Dolly Suite" and with whom Nan was playing the Secundo part of their duets. But by the time it came around to giving sixteen-year-old Allen his lesson, Nan's fatigue began to catch up with her. She felt as if her mind was racing but as if everything around her was in slow motion. She made a cup of strong tea to counteract the feeling, and powered through the lesson.

But then she knew she had to prepare for the trio's meeting that evening. She sat down at the piano, warmed up with some arpeggios, and ran through some passages of Beethoven's trio. She found she couldn't focus as she wanted to: somehow all she could think of was a stack of bills she had piled up on her windowsill and December's rent. Why *would* money intrude on music so frequently? How was she going to come up with enough to square all the minimum payments for her credit cards?

On a whim, Nan decided to call her former piano teacher, Lilian, and see how she was doing in Larkspur. Nan felt ashamed that she hadn't been to see her since before the pandemic.

"Nan?" Lilian picked up after the first ring.

"Lilian! How are you?"

"I have some visitors, you know."

"Oh, did I interrupt something?"

"Some rabbits have decided to pay a visit to my garden. I see them out there now, looking around for good greens for their dinner."

"Isn't that a problem for you as a gardener?"

"No. The way I see it, you plant an extra garden for the rabbits and a garden for yourself. That way, everyone is happy."

Lilian came from a town in Bavaria, and she had wonderful stories to tell of her explorations in the Black Forest during the 1950s and early '60s before her family had moved to Pittsburg. She was very much a free spirit. She'd met her husband Lewis in college at Carnegie Mellon University— he a gifted trombonist and she an even more gifted pianist. In the 1980s they had settled in the Twin Cities of Minneapolis and St. Paul, and Lilian had begun teaching piano. They'd bought a vacation home in Larkspur in the 1990s, where they had always gone to stay during the months of June to September every year. There Lilian's two sisters and their families lived (Lilian and Lewis had never had children). Six months before the pandemic hit, Lewis and Lilian, had retired to Larkspur.

Lilian asked, "When are we going to see you again? Things are better now with this crisis."

"Can I come see you on Friday, around eleven?"

"Of course! I will make my famous Schneeballen, as you remember."

Nan felt instantly better after speaking with Lilian, who was so much more than just her former piano teacher. Lilian had been a solid support the whole time Nan, Bess, and Teddy had been growing up, had been a good friend to Gwen, and had been the glue for an entire music community in Minneapolis for almost forty years. Lilian could be playful and theatrical, or she could be sober and philosophical, depending on the mood and the season. Nan loved her dearly in all her moods.

Nan suddenly had a resurgence of the enterprising spirit, and she shot off some emails to a few piano accompanists she knew, suggesting she could cover some of their shifts in the upcoming holiday season. She texted a few acquaintances for whom she'd sold things on eBay in the past and proposed doing so again. She cast about for anything she might sell from among her own possessions. She found two boxes of CDs that she had already burned into MP3 and MP4 files years before, deciding to check their quality, list them, and see what she could get for them. Finally, she texted a family who'd drawn on her dog walking services regularly in the past. She resisted the urge to go so far as offering herself as a substitute teacher again at some of the local high schools. While that job had sustained her off and on during the toughest of times over the last two years, she found it didn't combine well with teaching piano, composing, or keeping up her creative passions.

Nan had discovered, over the course of the pandemic, that though she was many things, she was no hustler: for her, the independence of being self-employed and drumming up her own students could never match the relief of having a steady income paid out by an institution. Most of her part-time enterprises were too sporadic and low-paying to add up to the kind of reliable income that would enable her to pay off her many debts or to be able to breathe more easily as she incurred daily expenses. She had come to accept the fact that she would probably always be behind the financial eight-ball, and that she might never be fully solvent.

At the trio practice in the Berkeley hills that evening, Jules and Céline were eager to discuss logistics for the performance of their trio in early December. Nan was so beat, she felt she would agree to pretty much anything. Céline reminded her, at the end of practice, that she was invited to dinner that Saturday at their home in El Cerrito. In theory, Nan looked forward to this. In reality, she felt increasingly beleaguered and frayed at the edges, and wondered where all the energy and time would come from to accomplish everything that needed doing. As she rode home on her bike, Nan tried to break down into sizeable chunks the things she had to do. She knew she had to plan out at least two recitals for her students, since the mid-year mark was fast approaching. She needed to consider carefully whom to ask to host these recitals and how to arrange the times so that everyone who wanted to could perform. At least Thanksgiving should be completely free, given the fact that the gym had canceled the boot camp for that evening. How she longed for a day in which nothing was booked! She wondered if Thomas would like to climb Mt. Diablo on Thanksgiving. She idly wondered what he was doing for the holiday. With Bess, Bertie, and Pooh-bah momentarily off limits to her, she had the feeling she might find herself alone. Margot always had a reunion with fellow fine arts graduate students in the City on Thanksgiving at the Wayfare Tavern.

The prospect of spending the holiday alone didn't seem so terrible. She could always volunteer to help serve the Thanksgiving meal up the street at the College Avenue Presbyterian Church. As long as she could fit in some music-making and exercise, the day would be well spent. Then there was the promise of playing Chopin at the retirement home's high tea the day before. As she took her various nighttime pills for sleep problems and bipolarity, she felt in a decidedly better frame of mind than she had in the afternoon. She would try to get ten hours' sleep, and all would be well.

The following morning, Nan missed her composing hours. She slept right up until 8:45, and had only enough time to eat a brief breakfast of leftover pancakes and prepare for the bike ride before heading out to meet Thomas at the base of the hill. She felt lighthearted and exuberant as she approached their meeting place. They hadn't gotten the chance to text much since Monday night, and she wondered how he was, what he was feeling, what he'd been up to. She marveled to think that in a mere eleven days' time she'd managed to begin falling so hard for someone. Where was he in this process? Had his admission of weakness been sincere the other night? She was inclined to believe him, as she was inclined to believe everything he'd told her.

As soon as Thomas arrived, Nan could tell he had an especial restlessness about him today. After briefly kissing her on the cheek, he led them at a good pace up the hill, as she struggled a bit to keep up. Thomas inveighed against the Belarussian and Polish governments, suggesting that the EU should penalize both countries for their handling of the migrants at the Belarussian borders.

She breathed heavily. "What's to be done about the migrants, though?"

"The response to this kind of crisis needs to be hammered out in advance at a worldwide summit, just like the climate crisis summits. It's no surprise, given the lack of concerted effort on the part of the world's nations to deal with climate change, that we should also be inconsistent in our ways of dealing with migration," he maintained. "But border control, like climate change, is going to be one of the major geopolitical problems of the rest of the century—if we make it that long on the planet."

"Don't you think the pandemic has basically shown that we're trying to crowd too many people onto the earth?" she suggested.

"Without a doubt," he agreed. "And beyond that basic cause, it's a pandemic born from too much mobility, on the one hand, and too much indoor living, on the other. Respiratory pathogens are bound to thrive under these circumstances."

"I hike sometimes with a girl from Kazakhstan," she said, still striving to match his punishing pace. "She says that the world is farming unhappy people."

He mulled that over. "You mean that as conditions enable people to reproduce more and survive longer, the quality of life is becoming steadily poorer?"

She nodded. "Yes, since cheap, mass-produced methods of providing for more people means that most of the world's population are being fed pap, literally and metaphorically. Masses of people are being conditioned to accept unhappiness as satisfactory—as more than satisfactory."

"As a teacher, you must view this problem primarily from the standpoint of education. Surely feeding students mere pap equates with encouraging them to have ideas and express them, but giving them no standards on which to base those ideas."

She was impressed with his acuteness. "Exactly. In my experience teaching at college and substitute teaching at local public high schools, very few students now are equipped to accept hearing the ideas of other people, because they haven't learned what it means to form an idea fully—that is, to have reasons backing it up."

"Do you think schools fail to teach debate?" he asked.

"They teach it, but allow very narrow ranges of viewpoints to be represented. For instance, one English teacher I substituted for was out for a whole week because the administration was 'reviewing' what had happened when he tried to have a critical discussion in class about the N-word as it was used in the novel *Invisible Man*. He referred to the word as it was used in Ellison's text, and got attacked immediately by four or five students who effectively shut down the discussion and reported it to the principal. The word was taken entirely out of context, and the whole staff turned against the teacher, treating him as an offender and the students as victims. He was still being forced to make apologies in restorative justice-type circles when I left to sub at another school. Ironically, ninety percent of his students seemed to be fine with the fact that he'd tried to open up a space for dialogue on the issue."

He spring-boarded from her story. "I think this is what a few Black academics like John McWhorter and Cornell West are speaking up against, calling it 'radical anti-racism.' They're complaining that teachers don't just teach historical facts to students, like slavery or Jim Crow, but they teach heavily biased academic theories that claim no real progress has been made against racism over the last century—that systemic racism is so prevalent, there's nothing anybody can do to successfully struggle against it. This kind of die-hard cynicism not only paralyzes the struggle against racism, but discourages intellectual tolerance and debate."

"And the fact that each side of the political spectrum loves to throw around phrases like 'critical race theory' and 'wokeness' without fully defining them only exacerbates the divide and slows down progress even more." She huffed as they crested the top of the hill, relieved that no more climbing remained. "Why so fast today?"

He looked surprised. "I think we're going the same average speed we usually do. My computer says 10.2 miles per hour."

"Oof, I must have overdone it with the weights yesterday. Or maybe I just haven't slept enough."

"We can slow up a bit," he offered.

"No, no. I'm glad you're keeping us honest," she said hastily. "Where to today?"

"Let's go down South Park, now that it's closed to cars."

"Okay." Nan secretly dreaded the steep descent that Thomas handled so elegantly, but which always had her white-knuckling the brakes.

He seemed in no mood for flirtatious back-and-forth today. Nor did he seem inclined to slow the pace so that they could breathe more easily as they talked. He was on a mission to thresh out ideas that current events had sparked in public debate, and his riding had a brutal efficiency that seemed to mock her relative slowness.

"You want to take El Toyonal on our way back up?" he said, when she'd caught up with him on Wildcat Canyon.

"Oy, you really do want to destroy our legs today! Okay," she said, thinking gratefully of the 34-34 gear ratio on the bike she'd had built two years before, when she'd moved out to Berkeley. Her bike had a white Merckx carbon-fiber frame with green, grey, and black detailing. She loved its all-round capacity for handling flats and steep hills, as well as its aggressive geometry.

There was always something so humbling and soul-testing about struggling up El Toyonal, one of the longest, steepest climbs in the Berkeley hills. Nan rarely put herself through this torture, but it seemed somehow fitting that Thomas should suggest it today.

When they'd regained the steam trains at last, out of breath and sweating through their winter layers, she asked, "Any plans for Thanksgiving?"

"Some friends are having me over in the evening," he said. "The bar will be closed, since all my employees want to be with their families. What about you?"

"No special plans as yet."

"Won't you and your sister get together?"

"We're kind of on the outs currently. In a bad way."

"I'm sorry to hear that. If you'd like to take a ride the day before, I'm free."

"That'd be great. Would you like to do Diablo?"

"The short version or the long one?" He grinned.

"You decide."

"All right, let's tackle South Gate and do the long version, coming down North Gate to the Walnut Creek BART."

Of course he would opt for that, she thought. They descended Claremont, Thomas, as usual, flying ahead of Nan.

"You're in a grueling mood today," she said, laughing, as they pulled over to say their goodbyes.

"Just feeling fit. We're pretty lucky to be as healthy as we are, don't you think?"

She had often had this thought—of just how easy it was to take fitness and general health for granted when you had them. "We are," she agreed. Then she added, smiling, "Are you getting a head start on giving thanks?"

"It's never too early to take a broader perspective on things," he reflected. "Or to appreciate the limited time we have."

"Hence South Park and El Toyonal?"

He arched his eyebrows, picking up on her playful tone. "We'll have to do those more often from now on."

♪ ♪ ♪ ♪

If Nan was disappointed at Thomas's aloof stance today, she didn't have much chance to think about it, as she got ready to ride the BART over to Margot's gallery. Margot had rented out a two-story loft space in a nineteenth-century building that took up the entire block on Rhode Island Street between 17th and Mariposa. Inside, she'd had a second-floor balcony constructed that ran all around the main room, after the model of the Guggenheim, which was one of her most beloved museums. Off from the balcony on either side of the gallery, a room opened up in which audio and interactive exhibits took place. Nan was in awe of how well Margot had

85

converted the space to a choreography exhibit where, months before, she had held a completely different exhibit on food.

"It's like an entirely new gallery!" she gushed. "I love the way you use natural light to feature each display."

Margot smiled. "If you can believe it, a lot of it isn't even natural light, but is a light most galleries use nowadays to mimic natural light."

As they walked through the exhibit together, visitors often came up to Margot to tell her how impressed they were with the works being shown.

Nan said, "I know I'm going to sound clueless, but how do you manage it when someone buys a work? Doesn't it steadily diminish the exhibit each time it happens?"

"The purchase is timed. I have two months before I need to hand over the work, and they have two months before they need to pay me. In that time, everyone gets to enjoy the full exhibit. See, there are dots next to the works that have been sold already." Margot gestured to an oil painting of some tango dancers, a pair of emerald-studded designer dance shoes, and a series of black-and-white photographs of dance shows on a cruise ship.

"How much money did you get for the tango dancers?"

Margot laughed. "You like that one? I do too. It went for $24,000."

Nan's jaw dropped. "Is it a famous artist or something?"

"An up-and-coming Marin-based painter who everyone is sure is going to be famous—which is just as good!"

"You must have to know a *lot* of people to pull off an exhibit like this successfully."

"All you need is to know a few of the right people, and they help connect you up with the artists, contributors, and investors," Margot claimed modestly. "As far as my own small team of technicians, assistants, and marketing people goes, I'm very, very lucky to have them."

Nan particularly enjoyed a series of dynamic sketches a Jamaican artist had done of dancers in the street in St. Ann's Bay, Jamaica. She also loved the film clips of flamenco dancers taken from a documentary currently being made about various dance troupes in Spain.

"Margot, this is out of this world," Nan exclaimed, as they finished the exhibit. "I sincerely wish I could afford to purchase at least a dozen of these works. Though I wouldn't have any place to put them in my apartment . . ."

Margot's eyes twinkled. "Why do you think I was so keen to own a gallery? This way I get to pretend for a few months that *I* have the space for all this great art."

After Margot had checked out with her assistant and taken care of a few emails, they left the gallery and went to her favorite taqueria up the street,

Dos Piñas, where they ordered Baja fish tacos, chili verde chicken fajitas, and two jarritos for drinks.

Margot led them to a quiet booth in the corner. "The beauty of this place—besides its ambience—is that it's open until 8, unlike most of the lunch-centered places around here."

Nan felt so relaxed when she was with Margot. It took her back to their graduate school days traipsing around Cambridge, Massachusetts, in which they often had no purpose to their rambles other than to talk for hours on end. Margot loved music, and they'd spent many memorable afternoons and evenings attending concerts at the Fogg Art Museum, Wally's Café Jazz Club in the South End, Jordan Hall, and various venues at the local colleges and music schools, where soloists and chamber groups performed.

"When are you seeing Aaron again?" Nan dipped a chip into the house-made guacamole, which exploded with flavors of cilantro and lime.

"Oh! I'm actually coming over to see him in Oakland on Saturday evening."

Nan pulled a face. "What time? I have a dinner with friends at 6."

"How about I text you afterwards, if it's not too late?"

"When is he coming to see the gallery?"

"I haven't asked him yet, but I'm hoping next week."

Margot shared some of the great travel stories Aaron had from an adulthood spent traveling and working through Europe, East Asia, and Latin America. "Needless to say, conversation never flags. He's very much at home in his own skin—more so than most guys I've met over the years."

They became suitably enrapt at the contrasting textures in the fish tacos and the fresh tenderness of the fajitas.

"What's new with Thomas?" Margot asked by and by.

Nan became pensive. "I mentioned to you we had what felt like a really passionate Monday evening at his place. And then today he seemed completely absorbed in current events and in pushing us hard on the ride. It could've been almost anyone he was riding with. I can't explain what's changed or what he's thinking."

"Let's see, none of my conjectures where Thomas is concerned have been accurate so far, but . . ." Margot trailed off, deep in thought. "Could he be stressed out about something? Like maybe work-related?"

"It's possible, I guess." Nan frowned. "It's true that I haven't exactly opened up to him about everything going on in my life."

"Oh!" Something occurred to Margot. "Does he have family coming in for Thanksgiving? That always puts people on edge."

"He said he's going over to a friend's house for the holiday, so I don't think so." Nan paused. "I hate to return to this, but do you think he's already decided we've gone far enough romantically, and he doesn't want to explore a relationship with me anymore?"

"But you said that the other night he confessed he was starting to have feelings for you," Margot pointed out. "That may actually be what's behind all this coolness. He may be trying to regain some distance because your relationship is becoming too all-absorbing for him emotionally, and he's freaking out."

Nan considered this. "Maybe."

"You also said he has a solitary side to him. Maybe this is him being a bit more withdrawn."

After they'd closed down the taqueria, they wandered back to Margot's studio in West SOMA. It was tiny but charming, and pretty much smack dab in the middle of the entire Civic Center area of downtown. Location had been everything for Margot in deciding where to live. Even more than Nan, she loved living in urban environments humming with the constant buzz of human activity.

They spent a lively hour playing with Romare the black cat, reminiscing about Cambridge, and talking about Nan's composition.

"It's a lot easier to work on it knowing I have the Meetup people to hold me accountable." Nan slowly trailed a ball of yarn on the floor, watching Romare's tail swish ominously and his yellow eyes laser in on his prey.

"That's exciting," said Margot. "So it sounds as if you're able to focus for now mostly on exercise and composing?"

Nan sighed. "I'm certainly trying to. What I still worry about, Margot, is easing up on the poetry, painting, and reading. What if I lose the techniques in drawing and watercolor that I've worked so hard on over the last twenty years? What if, by not regularly reading philosophy or memorizing poetry, I lose my facility with writing poetry, which depends so much on hearing echoes of other writers? I'm not so worried about keeping up the piano—I know I can bring up my music skills at any point with a specific goal in mind. But in fitness I already may be falling behind—look how Thomas nearly dusted me on today's ride." Nan laughed in self-reproach. "Sorry to spill like this."

Margot shook her head. "Don't be. No one lives in our own head but us, and it can be a lonely place."

"That's very poetic," said Nan, "and therefore true."

Chapter Six

On the BART ride back to the East Bay, a drunkard had pissed all over the floor of the car Nan happened to step into, and she didn't notice until, as the train sped up, the warm, powerful smell of the urine met her nostrils. The drunk man spoke to a few people, including Nan, all of whom completely ignored him. Meanwhile, his piss traveled in rivulets across various directions of the floor.

Nan had a sudden flashback to a similar moment in the New York subway a few years before. But she marvelled to think how differently New Yorkers had responded to the situation from the way Bay Area residents reacted.

It had been St. Patrick's Day of 2019, and Nan was returning on the 4 train from a concert on the Upper East Side. In her car, Nan observed how a porous drunk was reeling off a series of wrathful murmurings and hostile curses, regardless of the subway riders he addressed. She noticed that he had a pile of yellowish liquid beneath him. A polite young woman tried to reply to him, mistakenly thinking all people deserved a nice response. This was until she realized the dire situation, as potent whiffs reeked from the drunkard's mouth every time he growled in her direction. She eventually stood up, moved off a ways, and grabbed hold of a strap. Next, a friendly musician, a young man in his early twenties, not minding being engaged by the drunk, spoke with him about how Jimi Hendrix was the supreme guitarist of all time.

The drunk couldn't seem to decide whether to be a blithe drunk or an ornery, testy one. He lashed out at Nan every so often—who was a few seats away from him—but then immediately forgot he'd done so. There was no steady continuance to any one of his emotions from moment to moment. He threatened through various gestures to punch people, all the time muttering incoherently and slugging beer after beer. He was at first entirely chivalrous to all ladies who entered the car—"Wanna sit down?" he would ask slurringly. Little did they know what they were in for if they took him up on his offer.

Finally, he got out at a stop on the Lower East Side, and two beautiful tall women ventured to occupy the seats where the drunk had been sprawled. One of them proudly proclaimed to her companion, "This would be *far* from the first spilled beer I've put my feet into!" To which the other retorted, "*Or* the first puddle of pee either."

Nan laughed as she remembered this New York moment that showed how easily New Yorkers accommodated all the drab squalor of their City. The memory made her nostalgic for the Big Apple.

As the BART train re-emerged above ground at West Oakland, her phone brought in a text message from Thomas.

It read, "Are you already home?"

She typed back, "Am at West Oakland on BART."

He replied, "How tired are you?"

She wrote, "Surprisingly, not very."

"You want to get out at 19th and come to the bar? I have another hour or so here, but it's slowed up a bit."

Nan was suddenly glad she'd decided to wear her short lavender mini skirt and black racer-back shirt that evening. She'd meant the outfit to suit the stylishness of Margot and her gallery, but now it seemed useful in other ways.

Mixed with Nan's excitement was a little trepidation. She couldn't imagine a night with Thomas that didn't begin and end with questions, and this night was no exception. She pushed open the bar door and noted with secret satisfaction that every single face in the vicinity, male and female, turned to look her up and down approvingly. The bouncer Benny glanced briefly at her proof of vaccination and nodded her in. She looked around for Thomas, and saw him at a far table speaking to an older gentleman with a hawk nose and droopy eyelids. The music was lower tonight, and the atmosphere mellower. Thomas turned slightly and gave her an appraising once-over in which she perceived a flicker of lust. He said something to the older man, who rasped out a laugh, and then made his way over to where Nan stood.

"You look nice," he said, leaning in to give her a kiss on the cheek. Nan was aware that all eyes were on the two of them.

"Thanks. I just barely escaped skating on pee in the BART train."

Thomas laughed. "The delights of public transportation."

He gestured her over to the bar and nodded to two thirty-something Latino men, who paused their bustling activity to turn to Nan. "Elian, meet Nan. Nan, this is Elian. And this is Luiz. They're not just the bartenders—they're the backbone of this bar. Elian is from Havana, and Luiz from Rio de Janeiro. Nan comes from Minneapolis." Nan shook hands first with the shorter man, Elian, who had shoulder-length curly black hair pulled back into a hair band and pleasant, warm brown eyes. She then shook the hand of Luiz, who was tall and well-tanned with a buzz cut and deep-set green eyes. They both radiated charm and good humor, but even more strikingly, they had a genteel demeanor that showed they were a good fit for Thomas.

Thomas invited her to take a seat at the bar, and she selected a perch from which she could survey the entire place, apart from the people sitting in booths on the other side of a partition that ran up part of the middle of the room. This partition divided the bar into two sections, with the far section providing a small dance space at one end. Nan counted ten booths in all, five on each side of the partition. Towards the opposite wall, small tables were scattered, each with anywhere from two to four chairs. Along the middle of the bar, in line with but separated off from the partition, stood a double-fronted free-standing bar counter which could seat another sixteen people. The main bar, which was L-shaped, comfortably seated at least fourteen people. From these elements, Nan tried to work out the seating capacity for the bar, as well as its maximum capacity once standing patrons were counted.

"What's your estimate?" Thomas appeared at her elbow.

"Is it that obvious what I'm doing?" she asked.

"I like watching you think through it. It's very sexy."

"Okay, I figure about 122 seating capacity and . . . maybe another 80 standing, so a total of roughly 202?"

"You're not off by much on the seating—just figure, if you add a few more tables over there on the dance space you can fit in another 12 people. So the total there is 134. And standing capacity is actually 120."

Her eyes widened. "So you're saying 254 people can fit in this bar?"

"Easily," he said, smiling.

"You really *do* need to start serving food here as soon as possible. Think of all that money!"

"I do think of all that money." His eyes sparkled. "Especially now that things are better on the vaccine front."

He slipped behind the bar and began to mess around with bottles, an orange, and a barbecue lighter. She watched his nimble speed admiringly, not wishing to violate with words the almost priestly rites to which she was privy.

He set an old fashioned glass filled with brown liquor and a large square ice cube across the bar from her, and lit up a sliver of orange rind with the lighter, allowing it to blacken fully. He then neatly positioned the orange rind on top of the drink, which he placed in front of her.

"I hope you like this," he said. "It's inspired by a drink I had in the East Village when I last went to New York, in 2018."

She was surprised at the congruence of their thoughts that evening.

She held the drink up under her nose and breathed in the pungent aromas of peaty Scotch, smoky orange, honey, and some spice she couldn't recognize. "From its smell, I can tell this is going to be incredible." She took a small sip, and felt as if she was blowing small fires out of every pore. "Holy cow, this *is* incredible."

"I call it The Orange Dragon. Do you like the housemade habanero bitters?"

"That's what the spice is! Wow. I think this may be my new favorite drink. Is it all Scotch?"

"One-third Scotch, two-thirds rye. I use honey syrup to complement the tangy bitters."

"Where's yours?"

"I'm drinking beer tonight." He reached around to the other side of the bar and picked up a glass half full of amber liquid. They raised their drinks to one another, and sipped for awhile in silence.

"Who was the man you were talking to when I walked in?" she asked presently.

"That's Dan Fisher. One of the regulars who goes back to the early days of Harry's. In fact, he can tell you what this place was before Harry's. He's lived in this neighborhood more or less all of his life. I like to think of him as the Phil of our Cheers."

"Does he have a caustic wit?"

"He's certainly got a wry sense of humor," said Thomas. "And he's known for telling some risqué jokes. I apologize in advance if any of them ever cross the path of your ears."

"My mom always said that sort of thing builds character," she said.

"Is your mom no longer with us?" he asked.

"No. She died two and a half years ago. Of breast cancer."

"I'm very sorry."

"Thank you. But she was always very broad-minded. She always surprised me with just how open she was to anything and everything."

"My parents have both become much more tolerant since the four of us have left the house," Thomas observed. "When we were growing up, they could be very strict. Strangely, though, it was a different strictness from what other parents showed their kids. It was always based on rules my parents had devised for their own reasons."

"Did they explain those reasons to you and your siblings?"

"Yes, and the reasons made sense to us. For example, my parents would let us go to bed whenever we felt tired, or stay out relatively late on weekend nights, without asking any questions. But they held us to a strict account where our schoolwork and extracurriculars were concerned because, they said, this is the only job we have as kids, and you always need to do your job as well as you can before you knock off and play."

Nan nodded. "That does sound reasonable. What were some other ways they laid down the law?"

"All of us had to attend mass every Sunday until we were confirmed—not for the religious reasons other Catholic kids had to, but because, according to my parents, church gives a framing structure to the week, and until we could make a good argument for replacing it with something else, it was sacrosanct."

"What else?" She was riveted by his account.

"My dad was big on us keeping our promises. He said a person who doesn't follow through with a promise looks to all the world as if they're manipulating others, and no one wants to be around a manipulator."

"I like your dad," she said. "Are both your parents retired?"

"My dad retired from his dentistry practice just after the pandemic hit. My mom is still running her catering business, but she's thinking of selling it next year to one of her chefs."

"That explains a lot—you have cooking and hospitality in your genes. Where are your siblings?"

"Scattered across the Mid-Atlantic States—my oldest sister is in Philly, my next oldest sister is in Baltimore, and my older brother is in Brooklyn."

"So you were the rogue element that separated off from the pack—as well as the baby," She took another sip of her drink, her taste buds becoming accustomed to its spiciness. "How often do you see your family?"

"Pre-pandemic, every summer I'd take off a few weeks to visit New Jersey and as many of my siblings as I could manage. I'd also occasionally go out there for Christmas and stay with my brother and his family. I saw all of them this summer."

"And this Christmas?"

"I haven't yet decided. I may make a brief visit to Annie in Baltimore. She just moved with her husband and three-year-old son into a row house in Fells Point. In case you haven't noticed, my siblings and I are all drawn to more urban settings than we lived in growing up in the suburbs."

"Your house and neighborhood here strike a nice balance," she mused. "But do you ever miss the mayhem of Manhattan?"

"No question. There's no part of San Francisco that feels quite so frenetic or full of energy. But as I get older, I'm starting to appreciate having the best of both worlds—the frantic and the calm—in Oakland. And there's nothing like the cycling and hiking opportunities we have here."

"Am I keeping you from doing stuff you need to do, by the way?" She noticed that he'd begun to clean up a few things behind the bar. "Just ignore me if you have to."

"I can do most of this while we talk." When he turned away to line up some glasses, she admired his strong V-back and the sinuous outward curve running from his lower back to his firm, round cyclist's backside.

"What if Margot and I hadn't come here that night?" she asked teasingly. "Would you and I still just be meeting twice a week, cycling, and saying our goodbyes afterwards?"

"I was actually in a relationship for the first six months I knew you."

Nan was shocked. After a minute she managed to recover enough to ask, "How long were you both together?"

"We'd just started seeing each other when you and I met on the hill that day. So about six months."

Nan tried to check the growing jealousy she felt. "Was she a cyclist?"

"A spin instructor. She didn't like to ride outside very much. She loved the gym."

"Oh." Shaken, she still managed to ask, "But surely she must've been jealous that you and I were riding together all that time?"

"I was honest with her about you, and she knew it was important for me to have someone push me on the bike. She actually never really tended to get jealous. That may have been her mistake, or it may have been mine."

"What do you mean?"

"Remember how you and I were texting last week about the function of jealousy in relationships? I actually think a little jealousy can solidify the bond between two people."

"How could it have been your mistake?"

"In preferring things as I do, between two people." He gave her a meaningful look, and she surmised he was referring to his art of the long seduction.

"Where does jealousy fit into all of that?"

He nodded goodbye to a group of people who were heading out and who waved to him. "See you—have a good night." Then he turned his attention back to her, picking up on a more general thread of their conversation.

"Marina would never have been capable of or receptive to ideas the way you are. She wasn't nearly as articulate or brilliant or . . . witty. I never even thought of suggesting to her what you and I have spoken of."

Nan blushed to hear him heap such high praise on her.

He continued. "She had many good qualities that I won't go into. It's not the time or place. But it wasn't nearly as . . . compelling to be with her. We weren't right for each other. I realized that one day while we were riding and you said you wished you could keep in check some of your impulses to try to control everything. I said I wouldn't have pegged you as a control freak, and you said, half jokingly, that one of the things that kept you from being a control freak was that you sometimes loved losing control."

"I vaguely remember talking about that," said Nan. "That seems like ages ago."

Thomas smiled, folding his arms and leaning over the counter towards Nan. "I don't think I ever forgot any of the meatier conversations we had on our rides."

Faced with Thomas's distracting nearness and his charismatic aura, Nan fought to keep her composure. "But how is it that I thought you were single all that time? Did you give off mixed signals?"

"As I remember, you were pretty absorbed in finding jobs, making ends meet, and keeping up your different passions," said Thomas shrewdly. "You seemed pretty off limits at that time, even if I had been single."

Nan thought carefully about this. It was true that she and Thomas hadn't actually begun to flirt until around late May of this year. She hadn't thought of him as anyone more than a cycling partner until then. That was right about when she'd acquired enough students to be able to manage her finances a little better. How different he now seemed to her from back then!

"You still let a lot of time pass." Nan returned to the original point. "What if we hadn't decided to crash your bar?"

Thomas allowed his eyes, ever so slowly, to rove down to Nan's chest and then back up to her lips, where they settled for a long moment. "I guess we'll never know, will we?"

♪ ♪ ♪ ♪

As Thomas drove her home, Nan turned to him. "One more question—please?"

He regarded her with amusement. "I don't know which is funnier: the idea that I'm an oracle, or the farce that you would seek my permission to ask a question."

"Oh. I guess I have made rather bold with the questions lately."

He placed a hand on her bare thigh, which made instant heat radiate through her leg. "It's only what I expect of you."

She willed his hand to do more, but it remained in this semi-chaste position, as if teasing her with its indecision.

"You mentioned that it might've been your mistake in expecting jealousy from . . . Marina . . . because you, unlike her, have a taste for the long seduction. What role does jealousy play in the long seduction?"

"I think we have to admit that one of the main reasons this kind of seduction is for so few people out there is that it's largely in the head. Most people don't like being in their head so much. Part of being in your head a lot is having doubts about things—being skeptical of the world around you and second-guessing yourself. You suggested, in your text last week, that jealousy is all about doubting others and possessiveness is about doubting oneself. I would reverse that and say that jealousy is about doubting oneself. But that's what it means to be intellectual. I actually think jealousy is a higher form of possessiveness. Possessiveness is what all animals share. But to man alone comes jealousy."

"So it's not so much that jealousy is a necessary component of successful long-term relationships as that it's a *symptom* of them."

"A symptom that people read and then interpret as a sign that their relationship is working. So in that sense it's also an essential part of any relationship."

Nan turned to him. "Then jealousy is required not just in the long seduction, but in all relationships?"

"Yes, but I think it's required all the more intensely in the long seduction," said Thomas.

"You've really thought about this a lot," she marvelled.

The fingers of Thomas's hand began to trace subtle arcs around the inside of her thigh, causing her to melt into the back of her seat.

"I think I can guess your next question," he said, in a low, spellbinding voice. "The answer is no."

She bit her lip and clenched her fist, she was so focused on suppressing a moan. What he was doing right now was driving her wild, but she was determined not to give this away. The fact that he drove the car with perfect control while turning her on to this degree was all the more unsettling.

"How did you know what I was going to ask?" she said shakily. (Nan had wondered if he had ever carried out his plans for the long seduction with anyone before.)

"It's the logical follow-up to your last statement."

She let out a broken sigh. "All right, I give up. You've cornered me, in every sense of the word."

He drew up in front of her house, put the car in park, and placed his hands around the base of her head, drawing her in close as he leaned forward and pressed his lips to hers. Exposed as she felt, Nan let herself fall into the abyss that was Thomas's powerful hold on her. She arched her back into the kiss and groaned with pleasure. Her hand found his solid thigh and she used it as an anchor, pressing into it for support. The kiss lasted much longer than any of their previous kisses. Thomas by turns teased her lips with his tongue and then re-entered her mouth with equal masterliness, as if claiming what was his. Her entire body was vibrating now with the urgency of her desire.

He eased off, remaining an inch away from her face, as if struck with equal force by the extremity of the moment. She placed a hand on his heart, confirming with the gesture that his heartbeats were skipping as fast as hers.

"I don't know how much longer I can do this," she confessed. "It's destroying me."

He heaved a sigh that told of his frustration as well. "It isn't exactly easy from where I'm standing either."

In the back of her mind, she registered that he was making a Shakespearean pun, but she had no time for such distractions right now.

"You said the other night it's about what we want, and what makes us excited. But what about what we need? And about what's good for us? Don't we have to strike a balance? Just so that we don't self-destruct in the process?"

"It is important that we not self-destruct," he conceded.

She ran her index finger along his jaw line. "I sometimes feel these days as if I long for something safe, as well as something dangerous. I can't help but wonder what's on your mind when you step back from our intimacy as you did today. I fear you're trying to keep me guessing again, or that you've decided to put some distance between us so as to preserve your self-possession."

"Does this look like putting distance between us?" he asked.

"Not from this vantage, but earlier it did, yes," she admitted. "I guess what I'm saying is, that I do doubt you still—I doubt your intent, and your steadiness, and your own ability to trust. I can't help it."

"Let it be proof of my intent, steadiness, and ability to trust that I'm not offended by your doubt," he replied. "Rather, I see it as another promising sign of what's to come with our relationship."

"Are we in a relationship?" she asked, surprised.

"You doubt that too?"

"Everything is constantly in question with you."

"So I see," he said, smoothing out a furrow in her brow with his fingers.

"What if it becomes just me asking all the questions, and you stop having any for me? What then?"

He smiled. "You mean, what happens when we stop captivating one another?"

Nan nodded.

"To an improviser, does a well-loved tune ever get old?"

"How do you know about that?" Her surprise intensified.

He remained silent, awaiting her answer.

"No," she acknowledged. "There are an infinite number of variations you can make on such a tune."

"And it's always a delight to play the piece?"

"Yes."

"And to hear it?"

"Of course."

"I rest my case."

Chapter Seven

On Thursday morning, Nan forced herself to get up at 7:30. She was beginning to fear a relapse in her tendency to drift through the hours of the clock with sleep. But she was also determined to write every day from here on out, and to let nothing sidetrack her. She had a pretty tight line-up of gigs that day: at 10 she had therapy; at 11:30 she was meeting with a violinist she'd located through Craigslist, a college senior named Albert who wanted to play César Franck's gorgeous violin sonata; at 1:15 she was taking a dog for an hour-long walk up the hill; at 2:45 she was receiving some items from a friend of one of her piano students that she'd sell on eBay; from 3:30 to 5:30 she was holding lessons for two students at her place; and at 6 she had her boot camp.

She had an incredibly productive session composing, and didn't want to stop, but at 9:45 she headed out on her e-bike to meet with Naseer.

Today, Nan shared with him her frustration with the way she was always counting things and accounting for time.

"I remember when I used to be a lot more happy-go-lucky and easy-going," said Nan. "Sure, when I rotated around the clock with sleep I struggled with doing my job well and fitting in exercise. But my impulsiveness and free-wheeling attitude yielded a lot of unexpected encounters with the unknown and with people who were radically different from me. These are some of my most cherished memories. In New York I had a lot of late-night rambles in the West Village meandering from one jazz venue to another and meeting up with a diverse assortment of jazz aficionados and musicians. During graduate school in Cambridge I often let the moment carry me where it would after joining up with people at the gym, at concerts, in bars, or on the street, and our adventures stretched out into the early mornings. Now my careful meticulousness and tendency to control time down to the last minute seem to slow me down and to subtract from my existing time allotments rather than adding to them. Plus, I feel as if the high standards I've set for the use of time, including the time used for sleeping, have killed a sense of fun and joie de vivre that my old reckless self sported." Nan laughed wryly. "I even wrote a poem that begins, 'When all is meted out to perfection . . .' I know I'm an expert at counting, accounting for everything, and holding myself accountable. But it seems like a pretty high price to pay for normality."

At Naseer's sympathetic silence, she continued.

"Mom's death really accelerated this tendency in me. It reminded me of how short life is and of how little I accomplished before she died. I have a constant anxiety about not using present time well enough to look back later with satisfaction on what I've done now. But instead of counting out the seconds it takes to fill my spaghetti pot with water, I could be thinking of my composition. So I've shot myself in the foot."

"You may be putting a lot of pressure on yourself," Naseer cautioned. "Remember, we're always focusing on the short term and the here and now. How about for this week, you try to become aware of when you count, and substitute something else for it. What can you think of that's pleasant to bring to mind instead of numbers and time?"

Nan thought for a moment. "Being on the bike makes me happy. I'll think of riding in the hills. Minus the bike computer, of course—because that makes me conscious of speed, elevation gain, average speed, and so forth."

On her way back from her appointment, Nan stopped at her apartment and riffled through her collection of scores to find the piano part for the Franck. The last time she'd played this piece was in college, with a talented fellow music student named Sofia, who was one year ahead of her and who went on to play in the Mozart Orchestra under Claudio Abbado in Bologna.

Albert was a charming young man with a winning smile. He'd booked them a practice room in the UC campus' music building, Morrison Hall, and Nan was delighted with the tone of the Steinway in the room. In the course of their practice, Nan found that Albert was inclined to chat a lot, especially in between sections of a movement. They'd discovered they were both fellow Minnesotans—Albert hailed from the Minneapolis suburb of Minnetonka.

"Who's your favorite pianist these days?" asked Albert.

"Khatia Buniatishvili," she said without hesitation. "Although I always adore Simone Dinnerstein and venerate Martha Argerich."

"I saw Simone Dinnerstein play in New York when I was there this summer," he said. "I love Maxim Vengerov on violin."

Eventually, Nan warned Albert that she would have to leave around 1. While he was a gifted violinist, she coached him in the ways he could bring out gestures and phrasings in the piece to highlight its sensuality and variety of moods. They had only worked on the first movement by the time their session drew to a close.

"How many meetings do you think we'll need?" he asked cheerfully.

"For the whole piece? If you're willing to work on all the things we've discussed several times a week on your own, and if we meet for two hours each week—and really focus on the music—we could probably polish this into something performance-worthy by the third week of January."

He nodded. "Okay. I'll Venmo you the money later. Thanks so much for everything!"

Nan raced to pick up Hershey, the Australian shepherd mix that belonged to the Liu family a few blocks from Nan's apartment. She had a code for their house, and when she stepped in, Hershey went wild with excitement at the prospect of going on a walk. She was likewise overjoyed to see the buoyant Hershey again after a hiatus of a few months. They had a love-fest that lasted a good ten minutes, before she leashed him up, filled her backpack with water and dog treats, and headed out. They walked, in the crisp midday sun, up the hill a ways, passing Bess's place on the way. Nan stole a longing glance at the outside of her sister's apartment, as if hoping to catch some sign of where they stood in their sisterly row. She had hopes that as Oliver met more people in the area and became less lonely, and therefore less needy, Bess would come around to telling him the truth.

Hershey and Nan hiked up one side of the ridge of the hill and down the other side, which was steep and required careful negotiation around loose rocks and dirt. As they passed the elementary school yard on their way back, Nan scanned the groups of kids at recess to see if Bertie was among them. But these kids all seemed to be slightly older. She dropped Hershey off, texted a brief report of their hike to Calvin Liu, the father of the family and the one who'd hired her, and then rode home to make lunch.

She put some items on eBay, including her CDs, gave two lessons, and rode over to boot camp, where seven clients had come to sweat it out with her. By the time she got home she was, as she liked to borrow the term from the French, *crevée*—utterly done in. She ate dinner, played some Debussy on the piano, read some Wallace Stevens, and went to bed.

Grey skies loured early the following morning, and Nan noted the forecast that predicted rain. After composing for two hours, she loaded up her e-bike, put on a waterproof shell, and made the long ride up to Point Richmond, across the Richmond-San Rafael Bridge, and over into Marin County. On the way, the drizzle turned into full-on rain. Lilian's house was on the southwestern end of Larkspur, with immediate access, from its back garden, to a fire trail leading through the Baltimore Canyon Preserve towards the East Peak of Mt. Tamalpais. It was a beautiful two-bedroom ranch-style house built in the 1930s with lots of natural light and a large wood burning fireplace in the living room.

Nan kissed Lilian and Lewis, handing them a gift of an assortment of Bay Area wildlife potpourris and soaps that she had picked up at the Farmer's Market.

As Lilian invited her to divest herself of her outer gear, Nan asked, "Do you have an outlet I could use later to charge up my e-bike battery?"

Lewis showed Nan to an outlet, and she made a mental note to hook up her charger to it in a couple of hours. They all gathered in the kitchen to help Lilian bring out to the dining room the various items she'd made for lunch.

"Lilian, you shouldn't have gone to all this trouble," Nan chided.

Lilian smiled. "We haven't seen you in so long—this is a special occasion!"

"How are your nephews and nieces doing—and their children?" asked Nan.

Lilian gave her a full account of her two sisters and their children and grandchildren, who ranged in age from a newborn to a fifteen-year-old. Apparently none of the older children had taken to music, but all of them were athletic in some way.

Lilian had made a butternut squash bisque, Bavarian potato pancakes, pan-fried meatballs, and, for dessert, of course, her "snowballs," or Schneeballen, which were round, fried pastry balls dipped in powdered sugar. By the time they were on their second cups of tea, Nan felt utterly stuffed and longed to lie down on the couch and go to sleep. Rain was now coming down with a vengeance, which made being inside the house seem all the cozier.

"I wanted to show you my garden today," said Lilian with dismay, as they gazed out the windows. "You can just make out the purple stock over there, which has done well this year, and the snap dragons here by the door."

"Oh, that's what they're called—snap dragons. I see them a lot with my hiking group, and no one knew the name."

Lilian asked Nan all about her teaching, her chamber group, her accompanist gigs, and her composing.

"I really want to be composing all day every day," Nan confessed. "All of this other stuff seems like so many unnecessary distractions."

"I'm sure you'll find, though, that these distractions are feeding into your inspiration when you do write, and enriching the sound of the music you create," Lilian suggested.

"And they probably keep you sane," remarked Lewis. "Being alone all day with just your own thoughts can be unhealthy."

"Are you still sure you want to return to college teaching?" asked Lilian. "Don't you enjoy teaching students one on one?"

"I miss the subjects I used to be able to teach—composition, theory, appreciation, history. Teaching students piano, I hardly get to use any of the work I put in all those years for my doctorate. I miss the lively classroom discussions we had at Drummond, and the interesting questions students would bring to office hours. Above all, I miss the academic setting—being part of a larger institution with a library and a campus and colleagues."

"You are truly meant to teach at university then." Lilian nodded. Then she clapped her hands together. "Oh! We heard an astonishing concert of the Marin Symphony the other night. Brahms' first symphony and Schumann's piano concerto. Have you heard that pianist before, Orli Shaham?"

"No, I haven't," said Nan. "I'm glad to know you're both making the most of the re-opening of the music scene out here."

They spoke of Lewis and Lilian's plans to travel through Europe in the spring, pandemic conditions permitting, and then Nan, noting the time, apologized for the fact that she needed to take off.

"I have two students starting at 3:30, and I tend to ride a little more cautiously in the rain," she explained.

She made it only as far as San Quentin prison on her e-bike—the motor stopped dead a half-mile before the bridge. She tried to run a diagnostic using her phone. But she suspected something was off, because not only did the application's screen on her phone flicker on and off intermittently, but the charge level read as 90 percent, the same charge level she'd reached when charging the battery at Lilian's house. She'd traveled more than four miles since Lilian's house, so the charge should've been something like 85 percent. It'd been wet going, though, and Nan feared that because her front fender had broken off the previous spring and she hadn't yet replaced it, water had interfered with the battery or the motor.

She thought quickly. She had a little more than an hour to get back to her lesson. She didn't want to cancel on either of the students today, as they'd both missed their lesson the previous Friday, and the following Friday was a holiday so there would be no lessons. She hadn't thought to bring her wrench set, so she couldn't take off the front wheel, and would thus need to put the bike into an Uber SUV, if one was available. She selected this in the Uber app, but the nearest one was more than 50 minutes away. The rain seemed to mock her as it increased its rate, and she drew up her hood under her helmet. She wheeled the bike over to the meager shelter of a scrawny tree, and wracked her brain for the best way to handle the situation.

Nan glanced down at her phone and opened up the last text conversation she'd had with Thomas. Should she text him?

She hesitated, examining the last message he'd sent, late this morning: "Have a safe ride to your piano teacher's place." It was as if he'd intuited the dangers of this trip.

She took that as the cue to call him. He picked up after the first ring.

"Hey," he said.

"Hi," said Nan. "Are you enjoying this rain?"

He chuckled. "I just finished doing a workout. I'm getting ready to head to the bar. Are you still out in this weather?"

"Yeah. My e-bike stalled by the prison, just before the bridge. I don't know what's wrong with it, but I can't run a diagnostic. I don't have my wrenches, and there's no Uber SUV available within the next 50 minutes or so—this thing is too huge and heavy to fit in a regular car. I have lessons I really shouldn't cancel starting at 3:30."

Without missing a beat, he said, "Okay, I think we can get you back by 3:45, if you can move both lessons back a bit. If I leave now, out and back it should take a little over an hour."

"But how would we fit the bike in your car?"

"I have wrenches. We can take off the front wheel, put it in the back seat, and put the bike into the trunk. I may have to leave the trunk open, but I can put a tarp over the bike so it doesn't get any wetter."

She was flooded with relief. "Are you sure? You have to be at the bar . . ."

"I've got people to cover there until 4:30 or so. Text me your current location, and I should be there in under half an hour."

"See you soon. Thank you."

She called Neal and Charlee to see if they could move their lessons back 15 minutes, and then waited for Thomas.

He arrived in 25 minutes. "Traffic looks worse coming back over the bridge than it was going west," he said. By this time, Nan was so wet and bedraggled-looking, he instructed her to sit in the car while he unbolted the front wheel and hefted the rest of the bike into the trunk. In a few minutes, he had wrapped the bike with a tarp and secured the trunk into a half-open position with a bungee cord. He climbed back into the driver's seat and put the heating on low.

"Are you warm enough?" he asked.

"Yes, thanks. I'm sorry to have made you come out like this," she said. "I wish I knew what happened to the bike."

"The battery or the motor may have shorted out," he conjectured. "I'll be able to tell once we get it into a dry place and I can examine the contacts on the battery. They always say, check the simple stuff first."

"I think the rear end of your car is going to be dragging the whole trip back," she jested. "That bike weighs a ton."

Then she noticed for the first time that Thomas was wearing sweatpants, a tee-shirt, and sneakers, with a partially zipped-up hoodie.

"You literally just finished your workout."

He looked down distractedly. "Oh, yeah. I'll go back and change before work. Let's get you back."

Thomas tore at light speed across 580-East over the bridge, but, as usual, traffic slowed down to a crawl just before El Cerrito. Hitching car rides back with fellow hikers, Nan had often seen the freeway clogged this badly all the way to the exit to Highway 24, and it often meant waiting for half an hour or more.

"This is where being a racer comes in handy," he murmured, signaling and pushing his way deftly across four lanes of cars to get off at Buchanan. "I'm not asking you if I can come over; I'm telling you," he said, as if speaking to the various drivers that grudgingly made way for his BMW. From Buchanan, he angled in gradually by side streets that only street-savvy cyclists knew about, until he reached Sacramento Street, which was always open at that time of day going South. "The lights are usually with us on this street."

Nan was impressed. They made it to her place by 3:40.

"Let me take your bike to my place and I'll see what I can do. In case it is a short, I have a hair dryer," Thomas joked.

Nan briefly wondered why he had a hair dryer, but then pushed the jealous thought out of her mind. There was no time for that now. "I don't know how I can reciprocate . . ." she fumbled.

"I'll think of a way." Thomas's eyes sparkled as always. "You'd better get to your lesson."

♫ ♫ ♫ ♫

During her ride with Thomas on Saturday, Nan reflected on ways she and Bess could reach a common ground again. She decided to extend the olive branch once more. At their water stop, Nan texted her sister asking if she'd like to come over for some food at Nan's place on Thanksgiving. By the time Nan checked her phone again at the end of the ride, Bess had sent a message saying that she and Bertie had been invited to the house of a friend from work who lived in Danville, and did Nan want to join them. Nan texted back "I'd love to."

Thomas caught her smiling as she sent this message off. "Is that your sister?"

"Yes. It looks as if we may be able to make up after all."

"I'm glad for you. Fallings-out with family are never fun, but especially not around the holidays."

"Honestly, I don't think I could've stood living without Bertie for one more week."

"That reminds me," he said. "My friend Sue has invited us to a party the Friday night after Thanksgiving at her place in Piedmont Pines."

"Invited 'us'? As in, you shared with someone you know that there is an 'us'?" Nan couldn't help but laugh.

Thomas pretended to pull a face. "Ridiculous as the notion apparently is, yes."

"Is this friend so keen to see you partnered up that she'd go to great lengths to further the cause?"

Thomas considered this. "I've known Sue and her husband ever since moving to Oakland thirteen years ago. She probably would like it if I had a serious girlfriend again."

"That raises the stakes," said Nan mirthfully. "Thirteen years is a long time."

"So that's a yes?"

"How could anyone say no to a party in Piedmont Pines? Those houses are stunning."

"Twice insulted," he said, taking mock offense. "I'm glad to see that when you're in a good mood your first instinct is to take aim at me."

"That makes us a more plausible couple," she teased. "We have to look the part."

"If I'm to be your target in public, I can see I'll need to acquire a thicker skin." He rubbed gently at a grease stain Nan had on her forehead. "By the way, your bike is fixed. I can bring it over later."

"How did you manage it?"

"Water had corroded the lead points in the battery and shorted it out. I used alcohol and bike grease on the contacts and restored the circuit. The bike should run smoothly again."

"Now I'm doubly indebted to you. I'll have to make you dinner at my place. You said the bar is slower on Mondays . . . would Monday work for you?"

"Don't you have your Meetup?"

"I'll prepare a slow-cooked meal that we can enjoy afterwards. Come around 7:15."

"I look forward to it," he said.

When she returned from the ride, she decided to compose some more. Earlier in the morning she'd been in the zone, when she'd broken it to go meet Thomas. She was grateful to have more time to write this afternoon— now that he'd fixed her e-bike, she'd get to El Cerrito much faster for dinner with Céline and her family.

On his way to the bar, Thomas stopped at her place to drop off the bike. Both seemed to recognize that this was their last excuse to see each other for another two days, and they lingered a moment beside his car, standing a foot apart.

Nan felt the rush of a high, remarking buoyantly, "I think I've got a double obsession developing now: my quintet and you."

"That's high praise—to be considered on a par with your composition," Thomas replied. "Though I hope your finishing that project won't spell the end of me."

"You're strangely humble today." She smoothed her hand over the lapel of his leather jacket. "I like this mood as well as your others."

"You like that it renders me vulnerable," he said shrewdly.

"I'll admit, it's nice to know you have some diffidence."

He smiled wryly. "More than my fair share. For my part, I'll take self-assurance any day."

"Speaking of long-term projects, you must have a timeline for beginning to serve food at your bar?"

"If I can secure a chef recommended by Elian—a guy who used to work at a downtown eatery that closed during the pandemic—I can offer a bar-bites menu by the end of the year. I may be able to have a full menu in place by beginning of January. The trick is to hire and retain good staff. And the key to that is giving them competitive wages and treating them well."

"Will you have to do a lot of remodeling?"

"I've hired a contractor to put an addition onto the bar, and to install counters, prep tables, dishwashers, sinks, deep fryers, a grill, refrigerators, and another freezer." Thomas swiped a hand over his jaw with what Nan perceived as a hint of edginess. "The bar will shut down for eight days on December 15. During that time, the construction crew will be working on gutting the wall that leads out to our back courtyard—where all the dumpsters are now—framing the structure, putting in plumbing and electricity, adding the equipment, and decorating. We'll be open the 23rd and 24th, and then Christmas through the twenty-eighth the staff has off. We should have three lucrative days that conclude the year, and, if all goes well—a huge if—we can serve main courses by the first week of the New Year."

"Wow. So you've already done a walk-through with a designer and contractor. I had no idea it was this close to happening. If you need any help designing the menu, I'd love to give some input." Nan hazarded, "This must be nerve-racking for you as well as exciting."

"Believe it or not, the worst part is the name the bar has now. It comes down to what you said when we were at my place: 'The Tippler's Haunt' doesn't make sense anymore. But changing it involves a lot of advertising and putting out the word that we're the same place, only new and improved with food."

"Is it possible to get a head start on that now? By word of mouth, social media, Yelp, and so forth?"

He nodded. "I've been doing that for the last few weeks now—trying to create a little hype about our new set-up on Facebook and Instagram. I'm also designing a website with a spot for the new name left open. Any ideas you have are greatly appreciated."

"I'll think it over," she promised. "In the meantime, I'd better let you get to The Tippler's Haunt. Thanks again for the bike." She leaned in and kissed him, before stepping away.

As she watched him pull out from the curb, she remained astonished at the many dimensions one human character could have.

♪ ♪ ♪ ♪

Dinner with Céline, Arthur, Cécile, and Max was always full of delicious food, great wine, and lively conversation. Cécile had just turned 17 and, in applying to colleges, planned to major in microbiology; but she also wanted to keep music at the front and center of her life. Like her mother, she was a brilliant cellist; uniquely, she was a serious vocal soloist as well. She was applying early action to a series of Ivy League and other top-tier schools, and was in the process of crafting her personal essays. Nan admired Cécile's calm poise in the midst of an undertaking that would stress out most people. It was clear that the fifteen-year-old Max did indeed envy his sister's imminent departure to the East Coast and her upcoming freedom.

108

As Arthur handed dishes of cassoulet around the table, he said, "Nan, I've been to see your friend Margot's exhibit, and I find it very original. She not only has good ideas but good taste. By the time I got there, a majority of the works had been sold."

Céline nodded. "I'm wondering how she can sell that fresco of the tarantella dancers. How can they remove it from the wall?"

Arthur sipped his red wine. "I think the plaster layer is kept separate from a mount they use to place it on the wall." He returned to the gallery as a whole. "Margot has transformed that space into a different world. I used to pass that building from time to time, and it was a dump inside."

Nan glowed with pleasure. "Margot *is* sensationally talented. She's got many more ideas like this one in the works. I'll pass on to her how much you enjoyed her gallery." She turned to Max. "Speaking of turning things around dramatically, I hear you're largely responsible for the surprise success of your school's soccer team this year. You're a center forward?"

Max looked bashful. "Yeah."

Céline spoke up proudly. "He scored 21 goals this year, and those were against the schools with far stronger reputations for soccer. He's trying for scholarships at several serious away camps next summer in the Bay Area."

Cécile added, "When he plays with our cousins in Beaune, they always say he should aim to go pro."

"You all went back to Dijon this last August, right?" Nan relished the rich flavors of Céline's cassoulet.

Céline knit her brows. "That was only after our flights had been delayed and canceled four times. Covid has certainly made traveling more hellish than ever before."

"But it's all relative," Arthur pointed out. "Consider what our friend Marius went through when he was trying to fly back to the U.S. from Romania. The airline required a negative Covid test result taken no more than three days before travel, but he had to go three times on three separate days to get the test, because the results never came in within the allotted window. The Catholic Church and the nationalist politicians have a lot to answer for in deaths there. It's one of the worst affected countries in the EU over the course of the entire pandemic."

Nan pondered this. "As far as our own country's polarization over vaccination goes, I read an article suggesting that scientists try to establish common ground through moral values and traditions with people who reject pure reason and logic when making their decisions."

"How would that work if someone is trying to convince a religious person to get vaccinated?" asked Céline.

Nan buttered a piece of bread. "For instance, some Catholic leaders have called for their followers to see getting vaccinated as a moral obligation that demonstrates Christian love. They appeal to believers on the grounds of common sense and basic science."

"But it does come down to science and math," Arthur noted. "One can't get around the fact that some objective standards are needed to speak the same language in a debate."

"It's true that the idea of finding common ground through shared values and emotions works better in the case of, say, global warming," Nan conceded. "There an environmentalist can appeal to the climate-change denier on the basis of their shared love of nature or animals."

Arthur maintained, "Absent of politics, the ideal conversation would allow two sides not just to listen to one another, but freely to shift their points of view without feeling the burden of having to serve as a representative for a group."

"Yes, that's why tribalism and progress are mutually incompatible." Céline sighed. "Everyone stay seated. I'm just going to clear a few things for dessert."

Nan helped her take things into the kitchen and brought out the dishes for a grand finale: a Macaron Berry Sherry Trifle that Céline said was a French riff on English trifle.

"Who's the D.J. tonight?" asked Nan, who was loving the music that played in the background. "This is a beautiful playlist."

"I am," Cécile declared. "I made a mix of Spanish and Spanish-inspired music from the 19th through 21st centuries."

"Wow, that's impressive." Nan turned to her. "You are indeed a woman of many talents."

"You should speak." Céline laughed. "We look forward to hearing the début of your quintet in the new year!"

"Yes, and seeing what you make for your holiday watercolor card this year," added Arthur.

"Will you feature one of your poems on the inside, as you did last year?" asked Cécile.

"I hadn't thought of it, but I might." Nan was delighted at their enthusiasm.

"It was lovely last year," Cécile raved. "The image you wrote of the juniper tree blanketed in snow and the hollow cracklings in the cold air."

Nan laughed. "Now that I know I have such a receptive audience, I'm inspired to try my hand at it again."

“Do,” Arthur urged. “God knows most of the holiday cards we receive are dull enough.”

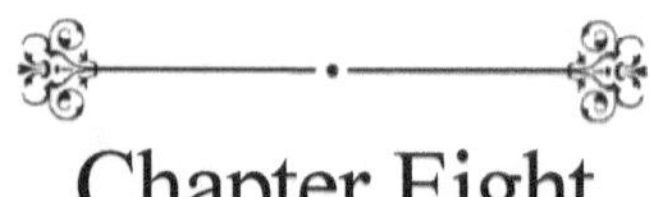

Chapter Eight

Nan was rather relieved when she received a text from Margot later saying that she was going to be staying too late with Aaron for her and Nan to meet up that night. Sunday was bound to be packed with intense mental and physical activity, and Nan was beginning to suspect she was burning the candle at both ends a little too frequently. It had been awhile since she'd felt refreshed after a night's sleep—she always could have slept in longer. According to her approximations from her sleep log, she had racked up a sleep debt of about sixteen hours now since the beginning of November.

After her succession of lessons had ended on Sunday evening, Nan called her dad. Frank was all ears as Nan told him how successfully the composing was going.

"That's great, darling," said Frank. "All the right elements must've conjoined this fall to enable you to launch into your writing this intensely."

Nan hesitated. "I've also started seeing someone."

"Oh! Who is he?"

"You remember the cyclist I've been riding with for almost a year now—Thomas?"

"Of course. I always wondered if something might develop between you two."

"You did?" Nan was startled at her dad's acuity.

"When you described him to me a few months ago, I heard attraction in your voice."

Nan was impressed. "Dads really do figure things out before anyone else."

She told Frank about Thomas's ambitions for the bar, his many abilities, his generosity, and his wit. She omitted the more intimate details of their courtship.

Frank listened intently. "You're going to miss him over the long holiday. We'll have to distract you with plenty of music-making, cooking, and walking."

They spoke of some books Gwen had sent to Nan a few years back which Nan was only now beginning to read: several novels by Barbara

Pym, Nancy Mitford, and M.F.K. Fisher. Nan had loved *Some Tame Gazelle* by Pym, and was planning on reading Mitford's *Don't Tell Alfred* next.

"Mom always said none of those books ever got old for her," Nan recalled fondly. "She said if she'd read a Pym novel once, she'd read it fourteen times, and the beauty of it was that she could dip into and out of it at any point in the narrative. She said when she woke up in the middle of the night and couldn't get back to sleep, she'd go into another room, so as not to disturb you, and read one of these books for an hour or so until she fell back asleep."

Frank seemed to hang on every word. "Yes, that sounds right. She always kept a stack of about a dozen books by the bed for just those purposes."

Nan asked her dad about Thanksgiving. "Where will you, Teddy, and Naomi be going?"

"We've been invited to the family next door. You remember the Carlsons. Normally Sarah and Ed have their kids home, but this year their kids are spending the holiday with friends' families in Seattle and Denver, where they're studying now. The Eckhardts, on the other side, will also join us— Denise and Glen, of course, never had any children. We're each bringing vegetable dishes. Teddy and Naomi will bring appetizers."

"Remember Mom's Thanksgiving dinner? That is still hands down the best I've ever had—or probably will ever have." This reminded Nan of the greater forthcoming celebration. "Teddy texted that the wedding preparations are afoot. He's put me in charge of the live music for the ceremony." Nan was very much touched by her brother's request. Apparently Naomi enthusiastically endorsed Nan's having a free hand with the music.

"I'm thankful that neither Teddy nor Naomi is fussy or materialistic in any way," said Frank. "Naomi is combing vintage clothing stores for her gown, and Teddy is sending out e-card invitations to anyone who's likely to be able to open them. They're discouraging guests from bringing or sending gifts. Of course, this won't stop some people from doing so anyways."

"How is Teddy handling the delicate matter of vaccinations?" asked Nan.

"He's not requiring proof of vaccination, but he's asked that those who are unvaccinated bring proof of a recent negative Covid test and stay masked during the ceremony and reception, keeping their distance from everyone else. He doesn't want anyone to feel as if their health is endangered by attending the wedding."

"That ought to be interesting with Auntie Jo's three—Emmaline, Lucy, and Curtis, and their kids," Nan remarked. "They're all anti-vaxxers, aren't they?"

Frank sighed. "Unfortunately, yes. It was with them in mind that Teddy came up with this policy."

After discussing the wedding a bit more, Nan and Frank said their goodbyes, Nan sending her love to Teddy and Naomi.

Nan made a shopping list of the things she'd need to buy at the butcher's shop and the grocery store the next morning for her *daube de boeuf à la Provençale*—beef casserole with wine and vegetables—from Julia Child's *French Chef Cookbook*. When Nan and her siblings were growing up, Gwen had cooked her way through practically every recipe in Child's various cookbooks, and to Nan this sort of dish constituted comfort food for the cold days of late autumn and winter. Best of all, it was meant to be cooked slowly in the oven for two hours, so she could put it in before her Meetup and finish it when Thomas arrived.

On Monday, Nan had timed everything down to the minute. After composing for two hours, she shopped, and did a twelve-mile run on the UC Berkeley Fire Trail. She prepared all the ingredients for the casserole, gave her two piano lessons, and, in between the first and the second lesson, put the casserole in the oven. She held her Meetup, which featured some of the same attendees as the previous week as well as a few new ones, and which Nan thought went very well.

Nan had set up the living room with a card table, her piano bench, and the only chair she had. She'd placed candles on the table, along with placemats and napkins, and had begun to boil the water for the egg noodles. She took out the Provençal fillip she'd prepared earlier, and put it on the table. She set her laptop to play the archived jazz programs from Sunday's radio line-up on KCSM. By the time Thomas arrived at 7:20, the apartment was redolent with the smells of beef, pork, and onions, and Louis Armstrong was playing Jelly Roll Morton's "Wild Man Blues" in clear tones over the speakers.

At the door, Nan enfolded Thomas in a warm embrace and kissed his cheek. He'd brought a bag full of red wines, having had advance word about the dinner menu.

"How are we ever going to drink all of that?" Nan admired his selection of Gigondas, Madiran, and Vacqueyras reds. "It seems you've opted for the southern Rhône and Pyrenees here."

"They're further West than Provence, but they're my own favorite A.O.C.s for French wines. They should go nicely with your casserole."

As Nan dropped the noodles in the water, Thomas stepped behind her and put his arms around her waist.

"You're looking particularly stunning tonight," he said.

"It's got to be the stained apron," she said, in an attempt to remain light. "That does it every time."

Thomas angled his head into the groove between her jaw line and shoulder, planting a deep kiss on her neck as he did so. She closed her eyes, shivering with delight.

"That might be part of it," said Thomas in a rich, enticing voice that vibrated through her body. "Or it might be that you've composed, given lessons and a Meetup, and also cooked a dinner for us—all in a day's work."

"Don't forget my run up the fire trail," she said laughingly, turning to him and meeting his lips with a kiss.

Thomas returned the kiss with fervor. Afterwards, he slowly skated the fingers of his right hand from the top of her forehead over the bridge of her nose, her lips and chin, down her neck, and all the way to the division of her breastplate, in one suggestive motion. She longed for his hand to cup one of her breasts at the end of its movement, but instead it slowly continued down to her belly, where it lingered a moment before settling again at the curve of her hips.

"Your hands seem to be a third guest at all our encounters. I admire their discipline." She traced her fingers over the bulging veins of Thomas's forearm. "How are these forearms not spoken for already?"

"I can see if I stand here long enough, my entire anatomy will be dissected," he murmured. "Personally, I'm in the mood to take in the whole, rather than the parts."

"And what do you discern in the whole?" she asked, removing the boiling noodles to a colander and straining them before running cold water over them for half a minute.

Thomas stepped back to let her finish the preparations. "A dauntless spirit, an even temper, a quick wit, and a low, calm voice."

"That's high praise." She flushed with pleasure. "But you still enumerated individual qualities."

He chuckled. "Then as for the whole, I see a precisionist."

"I can't quarrel with that." Nan placed the hot noodles into a bowl and tossed them with copious butter. Meanwhile, Thomas opened one of the wines and poured their glasses.

"Aren't we reversing the order of things tonight?" she asked gaily. "We're beginning with the foreplay instead of ending with it?"

"Words are our most potent foreplay. They begin and end every encounter we have."

Nan filled their dishes with casserole and placed them on the card table alongside two empty dishes. She then put the bowl of noodles on the table with a pair of tongs for them to serve up the noodles into the two extra dishes.

Thomas watched all this with the hint of a smile. "I see you don't like to mix the various parts of your meal."

"Almost never," said Nan with decision. "Thanksgiving dinner may be the only exception. This way you can carefully control the amount of casserole you have with each bite of noodles and vice versa."

Thomas nodded, as if this were the most natural line of reasoning. He took a seat on the piano bench, and Nan, after dimming the overhead light and lighting the candles, sat on the chair at the neighboring side of the table.

"Bon appétit." Nan raised her wine glass to Thomas's.

Nan was delighted to see that Thomas tackled the casserole and noodles with gusto. They were quiet for awhile as they ate to the sound of Bessie Smith singing "Preachin' the Blues" on the radio.

Nan asked, "What sort of food do you have in mind for the bar?"

"All the things my patrons are most likely to want," said Thomas. "American, Mexican, and Southern cuisines."

"So, burgers, tacos, ribs, fried chicken, that sort of thing?"

"And, of course, we'll throw in a few vegetarian options."

"Okay," said Nan. "That helps in brainstorming the new name for the place. I think of all the names you see on the eateries near Lake Merritt: Grand Lake Kitchen, Comal, Sidebar. They either announce directly what they're about food-wise or give an idea of their atmosphere."

"I place myself confidently in your hands," he said. "You're the wordsmith."

"I forget how much your place emphasizes draft beers? All I've had there is cocktails."

"We have about eight on tap, and a decent selection of bottled beers."

"Okay, so 'taproom' or 'pub' probably wouldn't feature in the name."

He smiled. "You're very methodical."

"What's something striking that comes to mind in your travels—in Austria, Hungary, and elsewhere?"

Thomas considered this. "In Vienna, Schönbrunn Palace, the Ferris wheel, the cathedral . . . In Hungary, the bars, castles, wines, Budapest itself . . ."

"'Budapest' might be a candidate," she mused. "After all, you do have Hungarian ancestry. Are you offering any Hungarian dishes on the menu?"

"I suppose I could offer one or two," he said. "With goulash being an obvious one."

"Too bad 'goulash' sounds slightly derogatory as a name. What about that famous Hungarian wine?"

He laughed. "There are several, but you're probably thinking of Tokaji."

"I guess that name suggests you're still all about drinking. What about 'Café Budapest'?"

"Then I really would need to offer some Hungarian items other than goulash—just to live up to the name."

"That's no good then. Hmm. Not far from you there's a place called 'Hopscotch.' What if you played on your love of cycling? What's your favorite mechanical part on the bike?"

"What flirtatious questions you ask," he remarked, quirking an eyebrow.

"Oh, I know. You're always riding without hands on the handlebars. What about 'No-Hands something' . . . " She considered this. "Somehow it doesn't have the necessary snap to it though."

She reflected for a moment. "In *Bicycling* magazine they're always asking riders on the last page questions like 'dawn or dusk,' 'pack or solo,' 'strength or grit.' I don't think I need to ask you, because I know already . . . but are you 'escape or pursuit'?"

"Pursuit," he said without hesitation.

"Yep, that's what I thought." She grinned. "So you could make the name of your place a pun on 'straight, no chaser,' and call it 'The Chaser Restaurant and Bar.'"

"I like that."

"You could even have a simple logo of two cyclists, one pursuing the other. I could draw up a few designs. The cycling slant would make an interesting note on your website and a conversation starter at the new restaurant."

He met her eyes and held their gaze appreciatively for a long moment. "Thank you."

She raised her glass. "To your new restaurant."

He clinked his glass against hers. "To the new restaurant."

♪ ♪ ♪ ♪

After dinner, Thomas caught sight of a small, travel-sized chess set Nan had on her bookshelf, and he convinced her to play a game of chess with him. She always took her time with her moves; none the less, the game lasted only eighteen moves before he had checkmated her.

"You have some creative ways of solving problems—both on and off the chess board," he observed.

"When I'm playing, I'm making up stories in my head about the various pieces," she confessed. "It's not all that useful, but it makes the game more fun for me."

"Speaking of making up stories, I'd love to hear you recite one of your poems."

She smiled. "Are you in the mood for something discursive and philosophical or more tightly woven and sensuous?"

"That's a tough choice." He took a sip of his wine. They had moved to the floor now, Thomas with his legs stretched out in front of him and his back leaned against the piano bench, and Nan in a side-sit position facing him at an angle. "How about both, in that order?"

"All right. You asked for it." She smiled. "This first is called 'Mere Company':

An old man always met them halfway,
Those anxious miracle seekers,
Showed them the spine of a seated cat,
Took them to the moonlit temple,
Bid them know the human form whole.

'All your life your story is perfected,
Like those folk songs of four hundred years.
Learn to tell it so your grandfather
And your grandson will listen,
Unfolding the mystery in personal ways.'

How has this boy slipped through my guidance?

The others know your plight, but not as well,
For you alone see what it is to love
Out of all mastering medium:
You are present in every drop of paint.

'Happy the man who teaches—who analyzes!
When would his cohorts brand him insane?
Born to manners he falls back on as friends.'

No, youth. At the artist's hands the critic suffers:
I teach you how to communicate
The cares I cannot feel but only speak.

Thomas kept his eyes lowered as she recited, and he continued to study a point in the near distance when she had concluded the poem. "This reminds me of folk songs in which the lyrics freely shift among points of view and time frames."

"That's what I think I was aiming for," she reflected. "Though honestly, I don't really remember—I wrote this so long ago."

"I especially like that line about how the critic suffers at the artist's hands—not the other way round. The second half of the poem makes me think of that famous Chinese saying, 'The student has become superior to his teacher.'"

Nan glowed with pleasure at his interpretation. They let his words sit for a moment.

"What do you have for me next?" he asked with a smile.

"This one is called 'Spring in Focus.' I wrote it in Cambridge:

Barbecue, box hedge, the bass notes of a guitar,
Retro wrap skirts holding day-old roses,
And the dilapidated corner with a sunken cart,
A bent bicycle tire, and an unclaimed sack—
Fugitive feel of folding day, the first
After an endless winter wrestling warmth—
We, roughly moulded, mocking our confinement,
Slighting our successes, calling back dim hopes,
Step forth audaciously pruned, but still attuned
To the real, the new that March wrote on and about us:
These notes, this smell, that feel, this cluster-portrait,
Point needle-thin depiction of distant riot,
Demonstration, thievery, business ventures,
And all the volatile mechanism of moving voices
Muted and quelled for many weary winters
Now discovering the meaning of time and action—
Opening up as we have focused in,
Have used a tired rose in form of fashion.

He frowned thoughtfully. "It's interesting how the contrast you draw between 'distant riot' and this focused moment of the poet could apply to many different contexts over the last century or so—even though you probably had something specific in mind. Spring can mean action and reflection at one and the same moment, even in one individual mind."

He paused a minute and then quoted, "'Have used a tired rose in form of fashion.' That's a nice line."

"Thanks."

"You prefer a slightly more formal style, I see."

"I admire so many modern and contemporary poets who can let loose and be freer in their verse. I haven't yet been able to break free in that way," she admitted. She didn't add that for the past three years she had struggled to conceive of the whole poem in the same harmonious way she'd done when writing the poems she recited for him. She'd deliberately picked two poems that she'd crafted during her years of graduate study and early years teaching at Drummond, which had been the heyday of her poetry writing and watercolors. She longed to break free of her unremitting obsession over details, which not only bogged her down in the creative process but destroyed the overall effect of each finished piece. "Desire" had been the first poem she hadn't belabored in some time.

"Isn't that the trajectory of most great poets' careers?" he suggested. "They start out imitating others' forms and then, in their later work, begin to define form in more idiosyncratic ways?"

She nodded. "It's unfortunate that I associate my poetry writing with procrastination and guilt over not writing more music."

"That should all be in the past now, though—since you *are* writing your music."

"It all feels way too tenuous," she said looking at his hand holding the wineglass stem. "A few false steps and everything explodes in my face."

"What do you mean?"

Nan hesitated a long moment. Then she came out with it. "What if I were to tell you that I'm bipolar?"

Thomas seemed outwardly unfazed, but his eyes darkened with concern. "Are you on medications?"

"Yes, but everything in life is carefully balanced, and if enough elements get out of whack at any point, there's a likelihood I can have a mood swing."

"How bad is one of your mood swings?"

"They used to be really bad in my twenties and when I lived in New York—but they always only lasted about a day or two. I'm bipolar-two, which means my episodes have always been shorter and less serious than bipolar-one episodes, but they've also been more frequent. Thankfully, since coming to Berkeley, I've had only one episode. I think the calmer lifestyle here has helped me maintain a better balance."

"What happens during one of your episodes?"

"So many things have happened. Neighbors in a nearby apartment called the police on me once in New York when I was screaming loud and long and crashing around at two in the morning. I wrote a three-page arrogant email dressing down an established professor in graduate school and he read it off to his colleagues in the department. I ended up humbly apologizing to him the next day—an apology he graciously accepted. I left a series of vicious voicemails for my landlord one time when the electrical outlets failed. In my Oxford flat, I deliberately dropped a tray of dishes and broke them all while stomping up the stairs and slamming all the doors, and then swore at a flatmate. In Cambridge I started punching trees and cursing people out who seemed to be ignoring the homeless people on the street. I spent over ten thousand dollars on clothes on a several-day-long shopping spree after a man I was falling for broke up with me. Once I even kicked out my sister and her boyfriend of the time at midnight when they were staying with me at my apartment in Somerville—I kept muttering, 'I can't help what I'm doing. I'm not in control of my actions.' One day soon after my mom died, back in Minneapolis, it seemed everyone in my family was asking me to do too much and I couldn't stop screaming, shouting, and playing crashing pieces on the piano. That's just the tip of the iceberg, I'm afraid."

"So basically you have an episode when you reach a critical level of stress that you can't deal with."

"That's about it. And all my twenties and early thirties I spent trying to get the right drugs and the right dosage to manage this bipolarity. If I took too much Depakote, for instance, I'd get sluggish, depressed, and suicidal, which I found far worse than the manic episodes. For instance, one time while living in Oxford, I contemplated throwing myself in front of a bus. Then if I took certain drugs to manage the mania, they made me too edgy. A psychiatrist out here has put me on some other medications that have helped a lot."

"But you still worry the scales could tip dramatically under the wrong conditions."

"Yes. For instance, not getting enough sleep is a real trigger. I need at least nine hours a night on average. When I get too deep in a sleep deficit I start hallucinating and screaming because I see spiders or hornets crawling up and down in front of me. This happened once during an Anne-Sophie Mutter concert. I was mortified. Consummate professional that she is, she kept on playing. It also happened when a friend was driving me to turn in papers at the end of term in my senior year at Brandeis. He had to stop the car because of my convulsions and screams. Those incidents were always linked to bipolar mood swings, or what I call 'breakdowns.'"

"What other pressures trigger the episodes?"

"If I feel obligated to satisfy too many different people at once, and it seems as if there's no wiggle room and no out, I can have a breakdown. Admittedly, a lot of the obligations are self-inflicted—I have to have a certain amount of exercise every day, time for piano practice, time to draw and paint, time to write and read poetry, and so forth. I need to feel I have control over all these things. If people and things start annoying me too much for the time they require, that can be a huge trigger. The stresses of living in Brooklyn were too much for me on a regular basis—the smells, the constant noise, the pollution, the crowdedness, the dirt. These made me suddenly flare up, and if I didn't curb my anger, it could turn into mania."

"So you can curb your mania?"

"I've gotten much better at doing that over the years, yes. If I never let myself begin cursing, for example, then I won't find myself escalating in my anger. If I'm around people, and animals, I care about—whom I don't want to have seeing me lose my cool—I can sometimes harness enough self-awareness to calm down. I've also come up with calming mantras to resort to at the start of a potential mood swing. And I try never to send emails or texts when I think I'm going manic."

Thomas had put down his wine at this new turn to the conversation, and devoted laser focus to Nan. She felt that now that she had begun telling him about this, she had to finish telling him everything.

"I should add that it isn't all bad. Yes, I've offended some people in the past, but I've always admitted my fault and apologized to them. My sort of mania hurts myself much more than it ever hurts others. It gives me great joy too."

"How so?"

"The bursts of energy are like nothing else. I've spent days on end playing through Bach, Beethoven, and Brahms on the piano—without hardly any sleep. I've composed whole volumes of poetry; I've painted some of my best watercolors; I've even had Olympic surges of athleticism in running, in the gym, and on the bike—all of this while I was manic. As far as I can tell, the output was legitimately excellent on all fronts. Admittedly, the feelings of grandiosity and superiority to everyone else— the feeling that my mind and senses alone could appreciate with heightened awareness what the universe offered—the feeling that my brain had sped up to the nth degree—all of this may have been somewhat exaggerated. But objectively, there was some truth to it. Because I was able to memorize whole speeches from Shakespeare and poems from my favorite authors in record time—and retain them. And when I recorded my piano or singing, it always measured up to my expectations."

Thomas nodded thoughtfully. "I think I know when you went manic here."

"Really?" She couldn't have been more taken aback.

"February of this year, right?"

"Yes!"

"You had a ride with me when you tore up the climbs and actually outstripped me on the descents you normally consider scary. The whole time all you wanted to talk about was how narrow-minded and nasty people were who were playing the 'mask police,' as I think you called them, in outside spaces where people were exercising at safe distances. I remember hearing you actually muttering the F-word at a lady who shouted out to us on our bikes from forty feet away to put on our masks. I laughed at the time, because it was pretty funny. I now see how out of character that was for you."

She nodded ruefully. "I pretty much only swear when I'm manic. I hope it doesn't ever happen again."

"That's what I was going to ask next. If you had your choice—to be in a state of equanimity or of mania—which would you choose?"

"I'd always choose to have, as you said earlier, 'an even temper.' I think I still create and perform well enough in that condition, and then I don't have to hide from those I love out of shame for my behavior."

"I don't think you should feel you have to do that anyway," he said, his brows knit. "Your bipolarity is separate from who you are essentially, and anyone who knows and loves you should understand that."

"That's a good thing to tell my moral conscience now," she said. "But don't let that on to my psyche, or I won't have a threat for it the next time I start to go manic."

"What can we—what can I—do to help keep you from having another mood swing?" He spoke with a softness she had only heard in his voice once before—on the night a week before when she had confessed that she was falling for him.

"Be you," she said without hesitation. "Be there. Be true."

He gave a crooked smile. "I think I can manage those."

Chapter Nine

The following day flew by, filled as it was with obligations small and large. Nan composed for almost three hours in the morning. Albert had opted to shift their Franck practice to today, given the upcoming Thanksgiving holiday, so they met on campus from 11 to 1. She walked Hershey again, and, because a few eBay sales had gone through, she had to pack up and ship the various items. She bit the bullet and called two of her piano students' parents to start arranging for the two mid-December recitals. She gave lessons to her students Joanna and Allen from 4 to 6:15, and then from 7 to 9 she met with the chamber group. When she got home, hungry as she was, she was wired and keen to sketch some drawings of cyclists in pursuit and escape for Thomas's restaurant. Over dinner she examined online pictures of track cyclists and roughed out a few schematic designs on the basis of what she found.

Nan was upset that she hadn't really exercised that day—apart from walking Hershey and using pedal-assist on her e-bike to reach various places. She was glad that Thomas had chosen to do the long route for Mt. Diablo the following day. This extra mileage and exertion should make up for her relatively easy Tuesday.

Thomas rode over to her house to pick her up on Wednesday morning. "It's only a little bit shorter to the base of Tunnel going the other way," he reasoned. "I might as well ride with you."

Thomas had a safe and enjoyable route planned out for them to reach the southern approach up Mt. Diablo. It was an absolutely glorious fall day. As they rode through the neighborhoods, Nan reveled in the blazing colors of the various leaving trees—birches, chestnuts, maples, persimmons, figs. She breathed in the intoxicating scents of coyote sagebrush, eucalyptus, pine, and bay laurel that wafted on the breeze in the hills. She admired the toyon berries dotting the canyon with bright red patches and the peeling, burnished bark of the madrones that contrasted so starkly with the dark greens and greys of brush and rock. Like the mounting for a gem, a high and translucent sky rose beyond on all sides.

"Do you ever feel as if the only time you'll ever really *notice* a place is when you first ride through it?" she asked. "As if every time after that the details become routine and you close your senses to them?"

"Yes, that rings true," he agreed. "I can't tell if it's because you tend to ride more slowly through a place the first time, or if it's because your brain is more highly attuned to it as something unfamiliar."

"It's strange that today, things we've ridden through so often look different—as if I'm seeing them for the first time."

"It may be you're anticipating the new part of the route that I'm showing you," he suggested.

"Perhaps," said Nan. But inwardly she had another thought. Riding with Thomas was starting to feel like uncharted territory the closer she got to him. To look at surrounding landscape now was to look at it from two sharply different perspectives—the period before she had been interested in Thomas romantically, and the period after. The first seemed like a well-worn groove; the second was like a fresh horizon. The contrast made her think of a famous ambiguous figure of a duck and a rabbit, where a person could see either image depending on the orientation of their vision. Her dad had whole books devoted to puzzling optical illusions and how they work, and, growing up, she had pored over these with fascination.

"Do you think, when we ride, we have a finite amount of sensory input we can process at any moment—so, if our thoughts are absorbed in the natural environment we relax our awareness of road conditions, cars, and so forth?" she asked.

"I don't think it's a zero-sum situation between the various parts of the brain, no," he replied. "The only proof I have of this is that there've been days when I've been sharply aware of everything on a ride—the weather, cars, cracks and glass in the road, the grade of hills, the landscape. Those days I've come home and discovered that the solution to a practical problem had come to me unconsciously while I was focusing so fully, as I thought, on the ride. It was as if my brain was working at full throttle."

She considered this. "I wonder if that's the converse of me. I think I'm a person of digressions and byways."

"What do you mean?"

"What most engages my heart and memory, and even my mind, is what comes to me incidentally while I'm absorbed in accomplishing something on the main, beaten path. Often I've skimmed through a text for an answer to a problem, but what I come away with from the text isn't the problem's solution—rather, it's all the material I've culled on the side while looking for the solution. And it's not the main points about art, music, philosophy, and so forth that I carry away from conversations—I remember the digressions. I justify this by saying that the side routes, the detours we take to get somewhere, are really the pith of life."

"So I focus on the byways and come up with a solution; whereas you focus on the solution and come up with byways." He smiled at her.

"My way isn't nearly as useful, I'm afraid." She met his smile.

They stopped for a brief coffee and sandwich at the Peet's in Danville and then headed East and North on Diablo Road. Nan could feel the magnetic pull of the mountain the closer they got to Mt. Diablo Scenic Boulevard and the climb up South Gate. She thought she understood why indigenous peoples had believed the devil resided in the four peaks of this mountain. It had a strangely mystical aspect that challenged and taunted, while inviting riders, hikers, and motorists to attempt it. She particularly loved the shifting views and micro-climates that met them as they climbed the switchbacks, the varied gradations of light and shadow that played across the mountainside, and the deep sense of history that imbued the lookout points and campgrounds.

They refilled their water and used the restrooms at the Junction Ranger Station, where they met a group of about a dozen cyclists who hailed from various parts of the world and the U.S.—Iran, Chile, Japan, Canada, New York, L.A.—and Nan chatted with some of them about the cycling opportunities in their cities of origin. It turned out they were part of a larger group of Bay Area riders that had organized this pre-Thanksgiving ride up Diablo. Some of them had never ridden to the Summit before, and they feared the steepness of the road. Those who had climbed it before, including Nan, shared stories about icy conditions during the winter, frigid high winds at the top, kindly drivers who'd given rides to cyclists who hadn't dressed appropriately, and wipe-outs from fallen rocks. After awhile, Thomas laid his hand gently on Nan's lower back, a gesture which had an electrifying effect on her body. She turned to him, and he said, "Shall we continue?"

"Sorry," she said as they began climbing again. "I easily lose my focus when people start telling stories."

"It's all right." He rode confidently less than a foot to her side. "I like the way everyone and everything interests you and sparks your curiosity. But I think you'll soon find we have little breath to spare on this ascent."

Sure enough, the last thousand feet of climbing tested their stamina and Nan found herself praying she wouldn't cramp up. She managed to make it up "the elbow," as people referred to the last fifty or so feet of intensely steep grade to the summit, and then, still out of breath, she began to pull out all the layers she had in her pockets.

"It's freezing up here, as usual," she remarked.

Thomas was also hastening to put on layers of fleece, nylon, and polyester—anything to break the wind's brutal onslaught on the descent down the mountain.

Once they had protected themselves from the cold, they headed over to one of the lookouts that faced down on the town of Clayton to the North and the Black Diamond Mines park to the Northeast. Nan ate a Power bar as they pointed out the peaks they could discern on Diablo. They then moved to the lookout on the opposite side of the summit, the one that faced Southwest and South.

"I think that's Mission Peak way over there." Nan pointed straight south. "At least, that's what one of my hiking friends, Larry, told us when we hiked up here this summer. Wow, it's really clear today. I wonder how many miles away that is."

Thomas put his arm around her waist and pulled her in closer.

"Am I too talkative today?"

"You do the talking; I'll do the fondling," he teased.

Nan loved the feel of his arm holding her, and she leaned into him, wishing she could put this moment on pause indefinitely.

"Who are the friends you'll be seeing tomorrow?" she asked.

"Beverly and Nick are co-owners of the company that distributes beverages to my bar. I've known them for nine years now. On past Thanksgivings, they've also had Beverly's parents over, as well as various friends and colleagues from the liquor supply business."

"You're going to get rip-roaring drunk, aren't you?" She fingered the zipper of his jacket.

"I confess I will be Ubering there and back." He smiled devilishly. "It's safer for everyone."

As if reading the longing in her eyes, he placed his hand under her chin and drew her lips towards his for a kiss. She closed her eyes, savoring the warmth and sweetness of his breath.

Presently he pulled back. "Should we start down?"

She nodded, already preserving the memory of this ride in the recesses of her heart.

The steep and twisty ride down North Gate was of exactly the kind of technical level that Thomas thrived on. Nan took her time, enjoying the vistas not afforded by the southern approach. When they reached the plateaus of the straightaway leading to the entrance of the park, they rode more closely together. As she often had before, she admired the shape of Thomas on the bike, the way his body looked perfectly positioned to handle whatever the road threw at him, the easy way he had of maneuvering around turns.

They rode in silence to the BART and waited in a sunny patch of the platform for the train to come. As always happened after riding down Diablo, Nan had become saturated with cold and she quaked violently. Thomas grabbed a pile of newspapers that had been left by a bench and opened up Nan's jacket from the bottom, pushing up the newspapers so that they covered her chest. "This is an old-school technique—but it works," he said.

"Th-th-thank you," she managed to stutter, her teeth still chattering.

Thomas wrapped his arms around her and held her firmly. "Just a few more minutes and the train will be here."

His body was warm, strong, and solid, and breathing in the fragrance of his sweat made her giddy with desire. Chilled as she was, Nan felt protected and thought she could have waited like this for another hour.

"You okay?" he said as the train pulled to a stop.

She nodded. "Much better. Thanks."

As she sat next to him in the train and recovered her body heat, Nan shyly examined Thomas's profile. She did this covertly as he took in the conditions of the car and the people in it, as he looked out the windows, and, after a few moments, as he became absorbed in thought and allowed his gaze to settle on a point in the middle distance. He had a striking profile—one that, to her mind, could command authority in any situation. She wondered what his employees saw in his profile, what his customers perceived in it, and what his friends tomorrow would see in it. Even if, until now, she herself could be blind to the forceful attraction and attractive forcefulness of Thomas's profile, surely there were many others who were far quicker to pick up on these qualities.

Nan silently berated herself for her slowness. She had missed out on six months of time with Thomas's profile! Nan checked herself mid-reflection: in the past she had built idols out of several boyfriends; she refused to do so now. It wasn't that Thomas didn't merit the idolatry—

for he more than deserved it; he had a kind of nobility to his spirit that kept her always humble in his presence. Rather, she wanted to avoid being his particular idolater. She perceived the strain that such a role would place on their delicate power imbalance. Like a teeter-totter that worked only when the children on either side were matched within a similar weight range, their rivalrous relationship could succeed only as long they remained close enough in strength and ability.

But she couldn't help imagining what the mind behind that profile might be capable of, and what the hands that rested on those thighs could do, were they ever to give her a display of their full capacity. She had seen both when they had been heavily curbed. What if their sensual expression were allowed free rein? She shivered with delight.

"Would you like my vest?" Thomas asked. "I'm warm enough now that I don't need it."

Nan blushed. "No, but thank you."

After they got off at Rockridge station, he rode with her to her place and saw her upstairs. "You really ought to have a hot bath at my place, but I know you have your retirement home tea that you need to get to."

Nan nodded regretfully. She too wished she could take a bath at Thomas's. Instead, her tiny cruise-ship shower awaited her with its water that never got hot enough and its broken showerhead.

"I'll see you Friday evening?" He stepped near her. "I think we should Uber up to Sue's place, just so we can drink freely."

"Thank you for such a lovely ride. I'll never forget today."

"What made it so memorable?"

"It was as if I was doing it for the first time," she replied. "I had a new perspective on everything."

"Remember, you only get that once," he reminded her with a smile. "I hope you enjoyed it while you could."

"I still am," she said. "Enjoying it, I mean."

♪ ♪ ♪ ♪

Nan took thorough pleasure in the high tea at the retirement home that afternoon. Fannie had arranged for sandwiches, cakes, scones, clotted cream, and jam to be brought in by a catering service. Nan played a steady stream of Chopin pieces on the piano—mazurkas, waltzes, nocturnes, ballades, and impromptus—and talk among residents and their guests was lively and interesting. Nan overheard Ethan and his family discussing internet dating—one of his guests referred to it as "meeting up through the machine"—and contrasting it with meeting up via church. It was Ethan who defended his grandniece's decision to use dating apps to try to find a long-term partner. An old friend of Maggie's had come over from San Francisco, and the two of them spoke fondly of people they'd known in their school days in Dublin.

Fannie made up a plate of goodies for Nan, which Nan gratefully accepted after playing for two hours straight. She was ravenous, particularly since she hadn't had time to eat much after getting back from her ride with Thomas. For the first time, she examined more closely the artwork on the walls of the common room. It was a hodgepodge of styles by different artists, some of whom were connected to the residents and others to the staff. One series of ink pen caricatures, clearly executed by a common hand, depicted various residents whose faces Nan recognized at once. She laughed to see the resemblances through their exaggerated features.

"You draw, don't you, hon?" Fannie came up beside her.

"Yes, and paint." Nan took a bite of scone spread with clotted cream and a dollop of jam.

"We'd love it if you contributed to the wall here. Anything will do."

"I'll look for something suitable." Nan was flattered. "Aren't your grandchildren going to be coming tomorrow—from Arizona?"

"That's right. It's going to be full at my house this weekend." Fannie sounded excited.

A resident in his early seventies, Ed, ambled over to them followed by a younger man, about Nan's age. "Nan, I'd like you to meet my son Dave. He teaches earth science at a high school in Benicia." Ed beamed proudly as Dave and Nan shook hands. "Nan here is a very talented gal, as you can see. She teaches music and *composes*."

Dave was over six feet, with wavy brown hair, a dimpling smile, and the merest hint of stubble defining a chiseled jaw line. Nan vaguely registered that he was what women often referred to as "strapping." "What sort of music do you compose?" he asked.

"Modern classical chamber music," said Nan. "I'm writing a piece for five instruments now."

"A quintet?" asked Dave.

"That's right."

Ed broke in. "Dave here plays the saxophone—he's really good!"

Dave reddened. "I don't really compose—I'm more into improvisation."

"That's wonderful," said Nan enthusiastically. "Improvisation's my weakest point in music. Do you play with a group?"

"Yes. We're called Echo Foxtrot 6, or EF6."

"Oh, that's clever—from the military code words. You picked the most musical terms in the lot. So there are six of you?"

"Yeah. A drummer, bassist, guitarist, pianist, and trumpet—and me."

"Judging from your name, I'm thinking you play classic and traditional jazz?"

"Mostly."

"Where can we hear you?"

"We play here sometimes," said Dave. "We also play at bars for open mic nights, and the occasional gig." He pulled his phone out of his pocket. "Let me take your number and I can text you some of the dates we're playing in the East Bay."

"That'd be great." She told him her number, he plugged it into his phone, and he shot her off a text in one swift movement. The text read, rather generically, "Hey, this is Dave." Nan texted back, fittingly as she thought, "Nan here."

With the forthrightness of an elderly person, Ed then announced, to no one in particular, "I've brought together two good-looking young folks today. I've done my kind deed for Thanksgiving."

Nan and Dave both blushed, and Nan turned to Fannie. "Well, I'd better be off now. It's going to be a long day tomorrow. Thank you so much for the delightful tea."

Embarrassingly, Dave walked her to the door. "It was nice to meet you," he said. "I'll write you about our gigs."

When Nan got home, all she longed to do was lie down and sleep. Instead, she sat down at the piano and ploddingly worked out some more fragments of her first movement. She had to use every last minute of time from now till the end of the year to compose this piece. How else would it get done? Before she knew it, several hours had passed.

Just before eleven, she received a text from Thomas. It read, "You up?"

"Yes, you still at the bar?"

"Just leaving."

"You want to have a nightcap at mine?" she shot back before she could overthink it.

"Be there in ten."

When Thomas arrived, Nan took in his fatigue, and gestured him in. "The long Diablo route plus eight hours' work spells disaster for Jack, you know."

At his questioning look, she clarified. "You know, the old proverb? Oh well, come in. I wish I had a nice sofa to offer, as you do. Or a fireplace, for that matter. Would you prefer a shot of absinthe or some herbal tea?"

"That's not a choice people get every day." He smiled. "I'll take the absinthe."

They stood in her kitchen, Thomas carrying the fragrant cold air from outside with him and Nan eagerly catching whiffs of his natural aromas in between the cold bursts.

"Getting a head start on tomorrow's revels?" she joked.

"It was a hard night at the bar, to be honest."

"How so?" Nan poured out their absinthe into tumblers. She ran the tap for a split second to cut each drink with a drop of water.

"There was a rowdy group of men in their late twenties. They seemed hell bent on destroying something. At last, Benny intercepted them as they were about to jump on the pool table in the back. He tackled one of them before he tried to split the table in two. Luckily, Benny has persuasive methods of getting people out of the bar when he wants them out." Thomas took a sip of absinthe, seeming to relish the taste of the licoricey drink.

"Do you ever have fights break out over women?"

"It has happened in the past. I've repressed those moments." He smiled wryly. "They're always the worst."

"I guess it's like what police say about domestic disputes—they're the most dangerous to intervene in."

"There's a particularly deadly kind of woman I've learned to look out for," he said grimly. "Outwardly she isn't as appealing as some others, but she has a kind of riveting singularity to her personality that draws certain men. She may be a Goth, for example, or she may be into vintage or grunge styles. Two men will latch onto her like bulldogs and fight it out to the death, as it seems. We've had some pretty raucous exchanges as a result of those women."

"Maybe once the restaurant opens, you'll have less of the rowdy atmosphere," she suggested.

"Maybe." He thoughtfully turned his tumbler around in his hand. "How was the tea?"

"It was good. I ate more scones and clotted cream in one hour than I have in a lifetime. I'm going to put one of my watercolors on the common room wall."

"That's excellent. I meant to ask you the other night if I could see some of your sketches and paintings."

She went to fetch her portfolio, as she playfully referred to her large folder of drawings and watercolors. She had filled it with her favorite items, very few of which dated past her mom's death.

Thomas took a seat on the piano bench, and Nan sat next to him, opening the folder. They moved slowly through each study and composition—a church in Venice; a series of historic buildings in Park Slope, Brooklyn; the profile of a boyfriend from ten years before; a dog rolling on his back; a fanciful depiction of goats, pigs, and other barnyard animals dancing around a fire at Christmas time; a still life of vegetables; an autumn landscape in Cambridge. He looked into her eyes with what she perceived as newfound admiration.

"You really are talented," he said. "These could sell, easily."

"I've never thought of doing art commercially," she said modestly. "I just do it to satisfy a longing to capture what I see in the world."

"Would you do a painting for the restaurant?"

"Of course. I'd be honored. What sort of subject would you like?"

"How about a still life like this?" Thomas held up the one of the vegetables.

"Would you like to have that one?"

"I couldn't take it." He put it away. "You made it with another purpose in mind."

"All right, as we make up your menu I'll compose a painting based on one of the dishes," she promised.

Thomas ran his left hand seductively along the outside of Nan's hip down to her mid-thigh. "That would be nice."

Nan tingled with anticipation of his next move.

"I look forward to seeing what you come up with," he added, leaning in and using his right hand to cup one of her breasts while planting a soft kiss on her neck, which he followed with a puff of cool breath.

She exhaled sharply, closing her eyes. "You're venturing into dangerous territory," she breathed. "I can't vouch for what happens next."

"Oh, yeah?" he murmured teasingly. "What have you got for me?"

She tilted her head up to the right, where he met her lips with his and enveloped them in an intense kiss. Somehow Nan felt Thomas's hands and mouth all around her then, engulfing her in their encroaching fire. Her thigh and breast burned with his touch. His desire was all-consuming, feeding on her sensitive areas—which seemed to spread all over her body.

"You have a sublime body," he whispered into her ear, as if intuiting her thoughts. "One it will be a pleasure to challenge and subdue."

She ached with longing, grasping his thigh in her hand and pulling his head closer to her. "Now?" she asked.

Thomas emitted a low laugh that aroused her to an even higher pitch of excitement. "Be careful what you ask for. Once you've been devoured, there's no undoing it." He skated his fingers delicately over the area of her groin and then settled his hand on her inner thigh.

She trembled from his touch, and her heart beat out a frenetic rhythm in her chest. "When?" she asked, and then, after a brief hesitation, "I want to be overpowered by you."

"I know you do," he said with a smile. Then he planted a slow, lingering kiss on her cheek. "We'll both know when the time is right."

"What if I say, as Orlando did, that I can live no longer by thinking?" she asked.

"If you can quote Shakespeare, it means you still have some facility to think," he said.

"Then I will deny you." She frowned.

"Deny *me*?" Thomas looked amused.

"Is that so preposterous a notion?"

"You have no power to deny me," he said lightheartedly, reinforcing his point by moving his hand slightly and causing her to groan. "We have gone beyond the point of no return. We have sealed our fates inextricably. We have bonded as artists committed to the beauty of our joint creation."

Nan sulked, tracing her finger over Thomas's taut abs. "What good is this creation if it annihilates us both?"

"You of all people know that we always have deeper reserves on which to draw than we think." He ran his fingers along the top of her scoopneck shirt, making her tremble. "You've said before that you've gotten your third wind as we ride."

"That may apply to cycling, but not to desire."

"All the more so to desire, which grows from lack of feeding."

"The desire may grow, but the mind may waste away."

"If I see signs of that, I'll concede to you. Until then, the more the mind contends, the greater the desire."

Nan reflected. "I'll grant you that discord may be at the heart of our desire. But if we dissent in this, as in other things, then how would we ever come to an accord where timing is concerned?"

"There may be struggle there, it's true," said Thomas. "Just as doubt exists in the deepest faith, or jealousy exists in the greatest love."

"Fine analogies, but who's to be the umpire ultimately?" She perceived her advantage.

"One or the other will have to capitulate. But it's in the nature of our dissension that we may both be surprised by who that person is and what form his or her surrender will take."

She couldn't help laughing. "You really do thrive on doubt. I begin to think you may be consistent in your philosophy after all—though it *is* a philosophy that's antithetical to security and harmony."

"You thrive on it as well," he asserted. "Otherwise this would never work between us."

"One more question."

"Of course." Thomas traced a delicate line along Nan's clavicle.

"If this is to be a literal *long* seduction, that implies that the first act of surrender—whenever and however it happens—is merely one, and the first, among many. So why should it be such an all-important event for us?"

"I think you explained the reason well on today's ride. You said one only gets to do something the first time once, and that's the time when one most enjoys it, and that's the time one best remembers. Am I getting that right?"

She shook her head, smiling. "You know you are, Socrates. I would've hated to go up against you in debate."

"You hold your own pretty well."

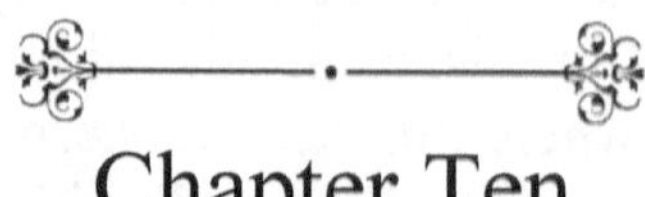

Chapter Ten

Bess and Bertie picked Nan up at 2 the following day in Bess's green Volvo station wagon. Next to Bertie in the back, Bess had piled a large salad bowl and two bags of vegetables she'd be mixing in the salad. Nan had made a butternut squash pie from scratch that morning—baking and scooping out the squash, after the manner of their mom—which she held in her lap in the passenger's seat alongside a bottle of wine. The two sisters kissed, and Bertie launched into an explanation of how Pokémon trading cards worked. Nan saw that in the brief eleven-day space since she'd last seen Bertie, he had picked up a new passion.

"Will there be other kids at the gathering today?" she asked Bess.

"Yes, some of my colleagues are bringing their kids. One of them, Benjamin, is seven, so Bertie should enjoy playing with him."

"Who's coming?"

"A couple of engineers—Adam and Scott—another stats person named Shaunelle, and a policy analyst, Dennis."

"Wow, will there be a lot of shop talk then?"

"I think we're all fried from the recent schedule, so I hope not," said Bess. "Janet, who's hosting along with her partner Terry, is a breath of fresh air because she's in accounting. Once projects go her direction, it means we're done with them for a while."

Janet and Terry's house was spacious and warm, and all the thickly layered smells of a turkey dinner in the making had settled comfortably in the air by the time Nan, Bess, and Bertie arrived. Bertie got immediately absorbed into a group of kids in the living room, and all the adults had flocked to the kitchen, which had a loft ceiling and opened directly onto an expansive deck where a fire pit had been laid with wood for later.

Bess greeted her colleagues one by one and introduced them to Nan. Adam and his wife Valerie, Benjamin's parents, had driven up from Sunnyvale and brought an assortment of cured hams from Spain and a keg of beer by way of before-dinner preliminaries. Scott and his wife Elaine had three of the kids in the front room, and Nan immediately noted their athleticism. Bess billed them both as dedicated triathletes. Shaunelle and her husband Grayson, who had the two youngest kids in the living room, had brought a delicious-looking spinach casserole made with baked yams and roasted walnuts. Bess introduced Dennis as "the new kid on the block," and when he smiled roguishly, she explained to Nan that two months ago he'd moved up from L.A. to join the team in Mountain View as their policy analyst. Something about Bess's flirtatious tone alerted Nan to take special note of Dennis. He was Black and a little over six feet tall, with a shaved head and high cheekbones, smooth caramel-colored skin, broad shoulders, and a warm, winning smile.

Terry handed Nan a glass of red wine and a small plate of hams and cheeses, and, after finding there was nothing immediately needing doing, Nan joined Bess and Dennis over by the deck doors.

Dennis was speaking about "the Great Resignation." "From the point of view of employment, anyone would say we're not in an economic crisis, because so many people are content to remain unemployed. The U.S. has the lowest job application rate in over fifty years, with four and a half million fewer people in the labor force than pre-pandemic. Workers just don't want to go back to work."

"Who can blame them?" Bess shrugged. "Everyone I know who's been able to work from home says they don't want to return to the kind of lifestyle where they're commuting three hours a day and working in the pressured environment of an office."

"And of course conservatives blame the government's financial support as a major reason so few people, comparatively, are applying for new jobs." Dennis took a swig of his beer. "But what's really happening here? What has the pandemic made people discover that they'll carry away for the long term?"

Nan sipped her wine. "Judging from the parents of my piano students, I'd say people have rediscovered family. More people in America want the kinds of social supports that European countries have."

"That, yes, and I think people have discovered their own mortality," Bess suggested. "So many of the workers whose jobs are the toughest to fill now are essential workers, and they're the ones hardest hit by the pandemic. They've seen and experienced the most illness and death. They feel more than ever the shortness of time in this life."

Dennis swapped out his beer for another. "Among our generation and the generation directly after us, I see people's newfound determination to be fulfilled in their career and to know their own worth, while avoiding mental and physical burnout. Older generations are calling us spoiled and entitled. I see this as only claiming our due."

Nan nodded. "I think the crisis is coming to a head in two particular areas—education and hospitality. Teachers and restaurant workers especially refuse to go back where they haven't been sufficiently valued or paid. I know from personal experience how hard the system has made it to teach in public schools without a credential, and how difficult, long, and costly it is to obtain a credential. Yet school districts have refused to relax their policies in the pandemic, and meanwhile, classrooms are overcrowded, classes are being cancelled, and a lot of districts have opted to move classes back online so they can hire qualified teachers from out of state. If the hospitality industry is similarly inhospitable to its workers, it's no wonder that restaurants and hotels are so short-staffed."

"My sister has a PhD in music from Harvard," Bess explained, "and she hasn't been able to get anything better than substitute teaching jobs in the schools here."

Dennis turned to Nan. "Unfortunately, I think there are all too many cases like yours of overqualified job-holders, on the one hand, and, empty jobs on the other. It reminds me of what Stalin said in the 1940s—apparently he announced that Russia had zero unemployment. When people looked into his reasoning, they found he said that because there were an equal number of jobs available as there were workers available to do those jobs."

"So it all squared on paper. Nice." Bess took a sip of her wine. "You had a lot of applications last month for your assistant position, didn't you?"

"Yeah." Dennis debonairly placed his hand in his pocket. "But that stands to reason. Every lawyer who's burned out in corporate law or litigation or who can't advance to a judgeship wants to try their hand at policy these days. But good policy jobs like mine are few and far between. The next best thing is to start out as an assistant. My assistant is vastly overqualified for the job."

"How long did you practice environmental law before switching to the non-profit sector?" asked Bess.

"Eight long years." Dennis smiled ruefully. "I wouldn't trade the experience, but it was hard going."

"You practiced in New York that whole time?" Bess took a bite of cured ham.

"That's right. When I realized I'd had enough, I followed up with some contacts from Stanford Law School and from some internships I'd done who eventually helped me land this job. At first I thought L.A. was the place to be. I now see that the Bay Area has a lot to offer."

Nan detected the playfulness that Dennis infused into this last statement, and she saw her sister smile in response. Over the course of the evening, Bess and Dennis joked a lot, sat next to each other at the table, and remained engrossed in each other's stories. Nan was delighted to see her sister enjoying the company of such an intelligent, charming, and handsome man.

Terry sat to Nan's right and proved to be an easy, down-to-earth person to talk to. She was a building contractor who'd been taught everything she knew by her father, from the age of six, in her home town of Louisville, Kentucky. After moving out to Sonoma in the late 90s and starting her own construction business, she'd met Janet at a bar in San Francisco sixteen years ago.

"We got this place thirteen years ago at an incredibly cheap price." Terry confided as if she knew her claim beggared belief. "It was just after the financial crisis, it had been foreclosed, and they couldn't offload the house fast enough. Of course, we redid most of it anyway."

"As I understand it, that's a far cry from the housing prices here now," Nan remarked. "I get the sense this is a very rich area."

"Yes, and unfortunately that corresponds to conservative political views among most residents," Terry conceded. "But we've managed to make it work. My business has done well here, and for Janet it takes under half an hour most days to get to the Institute."

Terry and Janet's black lab Queenie roamed in and out amongst guests' legs hoping she'd get lucky and score a piece of turkey or stuffing. The kids had their own table on the side of the dining room, and from it, eruptions of laughter and squeals of glee punctuated the adults' conversation at the main table.

Eventually, that point of dinner had been reached when the discussion became general, and when enough wine and beer had flowed that everyone was disposed to become expansive about a common topic the entire table shared. Adam and Valerie had apparently met through a mutual friend, and, as Nan gathered, the friend had been determined to break both of them of their "dating blind spots."

"What do you mean by 'dating blindspots'?" Elaine probed.

"A dating blindspot is a group of character types we can't ever see in an accurate light," Valerie explained. "Rationally, we might be able to locate and enumerate their faults. But we're never rational in our dealings with them."

"So each of us has a different dating blindspot?" Scott took a bite of salad.

"That's right," Valerie agreed. "In my case it was men who resembled my father, more or less—who critiqued everyone else for their lack of enterprise, but then ended up failing in every venture he ever attempted."

"And with me it was women who valued elitist institutions above everything else," said Adam.

"Our friend Charlotte identified this pattern in our dating and broke us of it by throwing us together." Valerie poured some more wine into her glass.

"What makes us irrational in dealing with these types?" asked Nan.

"Usually some deep-seated psychological reason," Valerie declared. "Or a sexual preference that's inextricably linked to our sense of who we are."

"Is a dating blindspot necessarily bad though?" Bess ventured. "I mean, if we're happy with the person we've chosen, who's to say the blindness is wrong?"

Adam nodded. "I think the blind spot is okay as long as it doesn't cause anyone harm."

"But surely a lot of us can't detect any real pattern like that in our dating?" Elaine countered. "Maybe some of us will never even come across a person who fits the type you describe."

"I agree," said Scott. "Our individual blindspot may be out there, but I don't think either Elaine or myself has had to go through that." They smiled at each other affectionately.

"I guess you're more likely to encounter this phenomenon the more you date." Valerie shrugged. "I think the aim is just to be aware of your blindspots and not become an unwilling victim to them."

Dennis spoke up. "My ex-wife was a dating blindspot for me. I may have been hers as well. Like much of my mom's family, my ex was dreamy, impractical, and increasingly inclined to be moralistic. As for myself, I was the athletic jock type my ex had always dated in high school and college. We married with the wrong expectations of each other, and couldn't ever really communicate as a result."

"How long were you married?" asked Janet.

"Six years," said Dennis. "We divorced three years ago."

"Dating blindspots probably don't work out in the long term because they start from a fabrication in the mind," Bess averred. "That may be a good basis for erotic love, but not for friendship, which is, I think, the main component of long-lasting relationships. For there to be friendship, two people need to have shared experiences together."

"That must be what Henry Adams meant when he said 'friends are born, not made,'" Nan remarked thoughtfully.

"Exactly," Bess agreed. "You can't *make* a successful relationship. It has to grow organically over time."

"So the element of time is the critical glue for a successful relationship?" Dennis poured more wine in her glass.

"That and maybe the element of grace too," said Bess. "At any rate, not a static, unchanging type that exists only in the individual mind."

"I was going to ask," Dennis continued. "What about those friends you meet for the first time and you can't explain why or how, but you know they're going to be your friends for life?"

"That's grace," Bess asserted. "Or, I guess in secular terms, graciousness."

"And friendship is gratifying." Dennis punned on her usage. "One friend makes the other feel honored, conferring dignity on him or her."

Bess smiled. "I think of friendship as the weekend of the worldly relationship's week. It's the time to be yourself without shame, to tell all, and let humor be king."

"Well, my good friends—" Janet opened her arms wide—"On that eloquent note, shall we repair to the fire pit outside?"

Nan watched her sister and Dennis take seats next to each other on one of the benches by the fire pit. She noted how attentive he was to her—how he refilled her glass and helped her with the sleeve of her jacket as she put it on. This was the first time, since Bertie had been born, that Nan had seen her sister show romantic interest in a man. Nan wanted to offer whatever she could to her sister to help make this relationship grow and flourish. Maybe Friday and Saturday nights she could babysit Bertie for Bess, and he could even stay over at her place. Maybe on Wednesdays when Bess wanted to go out after work, Nan could cook dinner for Bertie.

She glanced up at the dome of stars visible from Janet and Terry's deck, smelled the moist cedars in the yard, listened to the gay buzz of talk and laughter amongst the guests, felt the warm blaze of the fire, and tasted the tannic bitterness of her wine. She felt grateful on this Thanksgiving night to have her five senses intact and alert. She briefly reflected how much easier and more natural it had become—was it since she'd begun to fall for Thomas?—to open all her senses to the environment and savor the moment. Bertie came out, climbed in her lap, and soon began to fall asleep on her chest. As he slept, Nan's thoughts drifted to the man who was swiftly taking such a central role in her life. Had she conjured his perfections in her brain, or were they real? Which of his flaws was she blind to now, but soon to discover? Was she irrational where he was concerned? Did they have enough of time and friendship on their side to provide the basis for something enduring? In the midst of all these questions, Nan took comfort in the thought that both of them seemed to know themselves, at least. If one knew oneself well, surely that was the beginning of knowing another person?

♪ ♪ ♪ ♪

As if by silent agreement, neither Nan nor Bess spoke of Oliver that night. By the time Bess dropped Nan off, Bertie was sound asleep in the back seat. Nan wrapped her sister in a long and warm embrace. She sensed that Bess had distanced herself the slightest bit from Nan in the course of the last eleven days, but that essentially their closeness remained.

"I really like Dennis," said Nan, gathering her things. "Remember—any time, let me know, and I can take Bertie."

Bess laughed. "Wow, looks as if I should've started dating sooner. Okay, I may take you up on that."

It was only 9, so Nan decided to play some Schubert sonatas on the piano. All of her neighbors were young and childless, and all of them had enthusiastically endorsed Nan's playing after hours, especially on weekend nights.

As she played, Nan's thoughts, subconscious and conscious, were divided equally between her two obsessions—Thomas and her composition. She wondered if Thomas was the sort of person to become flirtatious and loose when he had drunk more than his usual share. Among those friends and colleagues from the liquor supply business who were at the gathering tonight, was there any single woman who right now was engaging him in witty banter, or coyly touching his arm, or glancing provocatively at him? Perhaps the woman was even stealthily grazing his butt as she passed behind him to refill her glass. For all this woman knew, Thomas was indeed single and available. And he was undeniably charismatic and arresting. Nan tried to shake off these jealous reflections.

Even in the midst of that thought sequence, she found herself focusing on the unique voicings of Schubert's chords, and wondering how she might borrow some of the more startling of these voicings for the purposes of her own piece. Perhaps because she had been playing so much Chopin the day before, a Chopin scherzo recurred to Nan's mind, in which a far-apart spacing of voices created a similar effect as the Schubert sonata she was playing now. Pausing the Schubert, she opened up the Chopin, took out her music notebook, and began to jot down ideas.

She didn't end up going to bed until 3 that night. She had had no texts from Thomas the entire day, even though she had sent him a text just before leaving with Bess and Bertie, saying, "Hope you enjoy dinner with your friends. What do you bring to a party of liquor suppliers?" She didn't worry overmuch about his silence, however. She knew he had a lot to organize with the contractor, and family members to call on the East coast. He may have been at the party past 11 and then been hesitant to text her, since they hadn't yet texted one another after 11. Obsessive as she had occasionally gotten in her own mind about some men in the past, she had never been a clingy type. Once a man showed the disposition to be independent and non-communicative, she gave him his space. On the one hand, she was far too proud to beg for what should've been freely given; on the other hand, she herself couldn't stand clinginess in another person.

That night, Nan dreamed that she was going through a reality she had experienced many times before (although later, when she awoke and remembered the dream, she didn't recognize it as a recurring dream). Within that dreamscape, she was a boy with a sister to take care of, and she could fly. Her sister could too, but most of the time her sister just held on behind Nan's back and let her carry her, which Nan was glad to do. They swooped high and wide over beautiful pine-wooded, cliffy, mountainous, gorge-filled landscapes. At night especially, Nan would venture out, feeling immune to all danger of human laws. She stripped off her sweatshirt and went cruising over an old house she remembered having visited many times before—that of a professor with a large family. By the time Nan reached the house, she began to wish she hadn't disposed of her warm clothes, because it was quite chilly. Nan stayed in the empty bedroom of a son who was away all night, perhaps partying, but for fear that he might come back, she didn't put on any of his clothes. This turned out to be a good thing, because he did eventually come back; they talked a bit, and when morning came, Nan felt homeless and an outlaw because she was loath to go. But she was worried once more about her sister.

When she awoke in the late morning, she remembered this dream and compared it to her swimming dreams. Her flying and swimming dreams were intensely satisfying for the sense of freedom they conveyed. For the remainder of the morning and early afternoon, she returned to the knotty second movement of her quintet. Around 2, a text came in from Thomas: "Shall I have my Uber swing by your place at 7?"

"That works," Nan texted back.

"Great—see you then. I have one Uber driver I call on so frequently that he lets me ride without a mask in the back seat. So feel free to keep your mask in your pocket."

She did a long resistance workout out back in which she incorporated a horizontal bar that hung over the picnic table by the barbecue grills. She alternated grueling sets of pull-ups, pyramid pushups, and pistol squats off a cement platform. She did walking lunges down the driveway with 25-pound dumbbells. Using an 80-pound kettlebell she did kettlebell swings and power high pulls. She did farmer's walks, bicep curls, tricep extensions, low squat walks with a resistance band, kettlebell snatches, and Romanian deadlifts. At the end of two and a half hours, she'd done twelve hard exercises that had targeted all of her muscle groups, and she'd sweat enough to drink two quarts of water.

After eating a light post-workout meal and reading for awhile, Nan scanned her wardrobe for the right outfit to wear that evening. Since they'd be inside a house the entire time, she figured she'd be able to get away with a white ruched racer-back mini-dress and high heels. She chose some gold-plated teardrop earrings with blue stones in the center that matched her blue eyes. She'd washed her hair, which was light brown in a shoulder-length cut that emphasized the heart shape of her jaw line and what her dad called her "graceful neck." By 7, she was ready for action.

"You look ravishing," said Thomas, as she climbed in beside him in the back of the Uber.

Nan took in his clean-shaven jaw, his casual but elegant maroon shirt that was unbuttoned slightly at the top, his open black jacket, and his masculine smell, and she tried to contain her giddiness. "You should speak. You're not so bad yourself."

They kissed, and the car shot them towards the Oakland hills.

"How was your Thanksgiving?" she asked.

"Which part of it? The first part was horrendous. The second partly made up for the first."

"Explain." She placed her hand on his leg.

He blew out a long, slow breath. "My oldest sister Ellie—the one who lives with her husband in Philly—is a hoarder with bad OCD. Ellie's friends, her husband Ken's friends, and my siblings all decided to stage an intervention this week. They persuaded her to see a psychiatrist and potentially get on medications, and they began to clean out Ellie and Ken's house. Honestly, I haven't even been there in over twelve years—apparently it's uninhabitable. Even the cat has gone insane from the lack of space."

"Wow. That's a lot to deal with. Is she going to be okay?"

"I really hope so. I guess the OCD has her alternating between anxiety and depression all the time, and she compulsively buys children's stuff and cookbooks. The first is probably vaguely intended for my brother Andrew's kids and Annie's kid. The second has to do with her obsession over food and nutrition. Ironically, Ken says that the kitchen is barely clear enough for anyone to make a cup of tea in, much less cook a meal in."

"When did they discover she had OCD?"

"About five years ago. But she didn't really want to do anything about it, and Ken let it go because the hoarding hadn't quite gotten to the point where he couldn't live with it. Now, it's gotten *way* beyond unlivable—Ken is worried the whole place is a fire hazard."

"I'm sorry for Ellie—and Ken. Do they have any kids themselves?"

"One son, Alex, in his mid-twenties. He teaches English at a high school in Shanghai."

"So that was your Thanksgiving. I'm sorry."

"Well, the get-together later at Beverly and Nick's was pretty awesome. Beverly had smoked the turkey thighs, oven-roasted the drumsticks and wings, and wrapped the breasts in foil and cooked them on a low heat for five hours, then glazed them with roasted garlic for ninety minutes at high heat. She is no-nonsense when it comes to preparing the various parts of the bird according to their individual taste.”

“She sounds like an amazing cook. Who else was there?”

“Bob and his girlfriend Amanda—a couple who run a distillery in Bayside, San Francisco. Grace and her sister Jackie, who inherited a gold mine from their father in the form of Townsend Liquors, a wholesale distributor that’s dealt with a large number of retailers in the Bay Area for the last forty years. An older couple, Pete and Irene, who own a chain of bars in the East Bay. And Sandy and Lou, Beverly’s parents from eastern Washington.”

“Did you all get tanked?”

“We did.” Thomas grinned rakishly. “I think you’d have enjoyed the dancing too.”

“There was dancing?” Nan tried to tamp down her jealousy.

“Yeah, Bob and Amanda were the DJs for the evening, and they had us moving even after we’d all gotten thoroughly stuffed on dinner. That’s saying a lot for their music selections.”

“Did you dance with anyone in particular?” she asked.

“Why do I feel as if that’s a weighted question?” He gave her a meaningful look.

“Maybe it is.” She traced her fingers over his quad muscles.

“If I did, the fact that I remembered so little this morning suggests it was nothing to write home about.”

“That gives me little comfort.” She kissed his neck, inhaling his alluring scent. “Were Grace or Jackie good-looking?”

“I guess they have the vampish good looks that go for beauty among a lot of men these days—heavily made up, rather anorexic, fake eyelashes, hair extensions, you know the type. They can’t be very pleasant to wake up to the next morning, when they can no longer hide behind their props.”

“Did they hit on you?”

“They may have. But no more aggressively than most women do at the bar. I’m sure my reservedness when faced with their glamour turns them off.”

“I’m sure of no such thing.” Her brow furrowed. “That can turn a lot of women on even more.”

Thomas turned and looked her steadily in the eyes. “I’m flattered by your jealousy, but you have nothing to fear where I’m concerned. You’re my type, and your type is so rare as to be represented almost solely by you.”

Nan was abashed by his frank tone, from which she perceived that he was playing no games and had no selfish motives. She berated herself for having doubted his steadfastness.

"If it's any consolation," he continued, playing with her earring, "I thought about you for most of the evening."

"So much so that you didn't text," she volleyed back playfully.

"They always say, drunken late-night texts are to be avoided at all costs." He smiled. "Oh, and to answer your text, I brought the makings for a potato, delicata squash, and arugula salad."

She laughed. "You have redeemed yourself."

♩ ♩ ♩ ♩

For the rest of the drive, Nan told Thomas about Bess's fledgling romance with Dennis and the conversations they'd had over dinner the previous night. Presently, the Uber pulled up in front of Sue's house in Piedmont Pines. The house had spectacular views of the Bay, and was built into the side of the mountain, so that it had four levels, the lowest fronting on a street directly below the street on which the highest level faced. There were about fifty people at the party by the time Nan and Thomas arrived, and an energetic, beautiful, dark-haired woman in her early fifties came over and kissed Thomas on the cheek. "I'm Sue," she said, extending her hand to Nan. "Welcome. Take off your jackets if you'd like, and come get something to drink."

A fire blazed in the fireplace, and Nan gave an appraising look over an area devoted entirely to musical instruments—a grand piano, a set of conga drums, a guitar, a double bass, and a flute. Thomas led her on a tour of the entire house, which Sue's husband Manuel had built himself. The lowest three floors shared the same loft ceiling, and two sets of stairs led up to various levels above, from which decks opened out with heating lamps turned on for those guests who wished to be outside. On the third level, a winding staircase led up to a fourth floor, where an immense skylight umbrellaed two bedrooms and a bathroom.

"This place is unbelievable," she marvelled.

"I know. Manuel is a really talented designer." Thomas placed his hand on the small of her back, leading her down to the lowest level, where they helped themselves to some red wine for Nan and some whiskey for Thomas. She loved the way he guided her in this subtle way, making her feel at once supremely feminine and cherished. They then joined Manuel himself, who was talking to a group of young men in their twenties and thirties.

One of the young men, whom Manuel introduced as Gareth, and who had a British accent, was explaining to a twenty-something man how we can know that time isn't just a construct of our mind.

"Think of a three-dimensional axis. If z didn't exist, we'd see everything flat." Gareth, crossed his hands vertically over one another to demonstrate. "We know time must be the fourth dimension because if time didn't exist, we'd see everything everywhere and nowhere at once."

Nan was immediately riveted. "But what about the relativity of time? Doesn't the rate at which time passes depend on your frame of reference?"

Gareth nodded. "Yes, but its relativity doesn't preclude its objective existence."

One of the other men spoke. "I'm Eric, by the way. One way my students enjoy proving that time exists is to say that you can order things and events so that they can't be put in any other order. Since someone looking on from afar can say *this* happened before *that*, and it happened *this far* before that, then not only does time exist, but it can be measured."

Another of the young men, who introduced himself as Vadish, added, "Putting aside quantum mechanics for the moment, a particle cannot be in two places at the same time."

"Exactly," said Gareth, "if you plot the particle at $t\text{-}1$ with respect to its position in space, and then you plot it at $t\text{-}2$, the difference between $t\text{-}2$ and $t\text{-}1$ will be t, which corresponds to a finite difference in space. Therefore, time must exist."

Nan countered, "But what if time flows and passes by, rather than being a fixed dimension like space?"

Eric had a keen glint in his eye. "A really good question. If we see time as constant change, our language, our propositions about the existence of things past, present, and future, and our notions of identity involve us in objective contradictions that none the less make sense from the standpoint of the individual consciousness."

"And all those contradictions can be resolved through the space-time continuum." Gareth took a definitive swig of his beer.

Manuel smiled hospitably. "Nan and Thomas, Gareth teaches physics at Cal. Mike is a student in sociology. Vadish is getting his doctorate in engineering. Eric is a lecturer in the philosophy department. And Matt teaches math at Cal State East Bay. So you can see how we came to be discussing time."

Matt turned to Nan. "What do you do?"

"I teach music. I used to be a lecturer in music at Drummond University. Now I'm teaching piano and trying to find another academic position out here in the Bay Area."

Matt looked thoughtful. "I nearly majored in music. But my parents won out in the end, and music gradually became just a hobby. What's your favorite subject to teach, within music?"

"Music history. Mostly because my dissertation took a historical approach."

Vadish asked her to elaborate, and she explained that she'd studied troubadours and minnesingers and the texts to which they'd set their music.

Matt's eyebrows rose. "Surely there can't be many troubadour songs that survived?"

"That's right," Nan affirmed. "Only some 300 out of about 2500. But my interest was equal parts literary and musical so I had plenty of material to mine." She gestured over towards the area by the piano. "Someone here apparently plays quite a few instruments."

Gareth said, "Oh, that's Alice and her students." He looked around as if to find Alice. "Sue and Manuel let Alice give her music lessons here. I've taken drum lessons from her."

Nan turned to Manuel. "That's wonderful. Do you ever hold concerts here? This is such a fantastic space."

Manuel nodded. "Wait till later, and we'll probably have one tonight."

As Gareth, Vadish, and Eric got into a discussion about UC Berkeley's recruitment for their respective departments, and Mike and Matt sought out more beer, a young woman drew Manuel away. Thomas smiled at Nan and leaned in to murmur, "Now it's my turn to be jealous. That's a lot of intellectual testosterone directed at one woman all at once."

Nan blushed. "I find the time question particularly fascinating, because I always think that one definition of brilliance is an understanding of the fleetingness of time."

"It's one thing to understand it and another to live by it." Thomas took a sip of his whiskey. "The real brilliance must consist in being able to do both."

Two couples had joined them and were listening intently. Nan and Thomas briefly introduced themselves and continued.

Nan said, "You're right. And as Eric hinted just now, people appreciate the transitoriness of things in different ways, so this kind of brilliance is relative. One person might translate it into savoring all the minutiae of life, while another person might direct their efforts towards worldly ambitions."

The tall blonde woman, who'd introduced herself as Claudia, chimed in. "That rings true. The French believe that the bulk of love consists in the little acts of sharing things together. Who's to say that the more dramatic affairs involve greater love?"

Sam, who was the partner of the other woman, tipped back his beer. "I think the first kind of brilliance is private and domestic, whereas the other kind is political."

Thomas spoke. "I see the first as a comic brilliance, and the second as heroic brilliance. Comedy is all in the details. Heroic drama is about the bigger picture. But the comic kind is in no way inferior to the heroic."

Nan asked, "Where does social brilliance fit into that scheme? For example, the decision to marry or to have kids is not merely a trivial, private one; nor is it a political one, usually. It falls somewhere in the grey area where the domestic and social spheres overlap."

Claudia nodded. "My brother-in-law insisted on having as many kids as possible, even though my sister would gladly have stopped at three. For him, your greatest accomplishments in this world are your children."

The other woman, Tamar, observed drily, "The brilliance, in his case, would involve sufficiently supporting those children and raising them to be good citizens."

Thomas laughed. "For other sorts of men, brilliance consists in dodging both marriage and children."

Claudia snorted. "Such men cannot be considered brilliant. They're just ignoring the cogs that keep society moving and functioning."

Nan nudged them back to her starting point. "Of course, it depends on cultural context. But in most cultures still, marriage is an important way to establish positions in society—it's not just a financial transaction."

Sam replied to Thomas. "And dodging children, for many men, simply means dodging their responsibilities towards the children they end up having."

Claudia's partner, Greg, remarked, "But I've known men who've carefully thought through the problem of having children and decided they don't want to bring them into the world—for various reasons. Sometimes it's because they're afraid they'll be bad parents. Other times it's because they look at the god-awful state of this world and the planet, and shudder to think of introducing more life into it."

Thomas turned partially to Nan. "Perhaps social brilliance consists in adapting your life to suit your needs within the framework of your own society's conventions."

"Can you give an example?" asked Tamar.

"Well, I think of my brother." Thomas addressed the group. "He and my sister-in-law took their time—seven years, to be precise—before they finally decided to get married. They lived together and even had two kids before they married."

"They sound very European," Tamar commented. "Why did they wait so long to get married?"

"My brother wanted to build up his wealth management company first. His wife was similarly focused on her career. But they didn't want to wait too long to have kids. So they opted to do things slightly out of order." Thomas finished his whiskey. "In the end, they married for financial and social reasons combined."

Nan observed, "It sounds as if both your brother and sister-in-law have a keen awareness of time's swiftness."

"Speaking of which, it's time we got something to eat." Greg smiled. He and Claudia excused themselves to head towards the tables on which the food had been laid.

Thomas turned to Nan. "Shall we also forage a little? You must be hungry."

Nan was famished. "Yes, let's."

After filling their plates with various tapas, crudités, and mezes, they climbed up to the deck on the third level, where the most breathtaking views could be had. Thomas had moved on to drinking mezcal, and Nan had filled her glass with a different red wine.

Nan munched on a radish dipped in hummus. "These days I'm feeling singularly un-brilliant. And that's despite working harder on my composing than I ever have. I feel as if I could've been writing like this all along, and I would've saved a huge amount of time in the long run."

Thomas chewed on an olive. "You can't second-guess yourself, though. Maybe the time has never been right before now. What about my own situation? I could've turned my bar into a restaurant long before this. I only really wanted to make it happen when the pandemic hit and closed everything down." He smiled. "You and I are both more focused on the comic aspects of life, I think."

"I'm glad you defended that. And, by the way, I'm quite happy with the contradictions that time affords. I don't feel I need to resolve them through the space-time continuum."

"All the same,"—Thomas swiped a grilled calamari from Nan's plate—"I'm glad people know about the fourth dimension. Objectively speaking, it's better for things to exist absolutely."

She laughed. "Yes, I suppose so. Getting back to appreciating time, though, it's not that I don't—I've always tried to get the most out of every moment. But I'm unwilling to compromise on so many things. I'm reminded of that expression, 'Don't let the perfect be the enemy of the good.'"

"So you think if you could give a little on some things, and remain satisfied with only being good, and not perfect, that you'd feel more fulfilled in the long run?"

"Something like that." Nan sipped her wine.

"What if you conceived of your pursuit of excellence in similar terms to the way I think of our long seduction?" The guests at the table ten feet away were loud enough in their discussion that Thomas didn't have to lower his voice to be discrete.

"How do you mean?"

"There's not really a finite endpoint to it. It's always a work in progress that can be perfected over the long term and adapted to changing circumstances. Maybe right now you have a high demand for finishing your composition. So you have to back off a little on your other interests. But you'll make up for it over time."

"You're not suggesting we back off a little in our relationship, are you?"

Thomas hesitated. "I'm saying we could, and it wouldn't affect the outcome in the big scheme of things."

"I disagree," said Nan. "In both areas you build up a momentum and have to keep it going. The pressure is part of what makes both performance and lovemaking worthwhile."

Thomas crooked his mouth. "A certain amount of pressure, no doubt. But you wouldn't want too much in either field, or you'd crack."

Nan smiled momentarily at his double entendre, and then resumed a straight face. "But could you really slow things down in what you and I have going right now?"

"If it meant you could preserve your sanity and be able to accomplish what you need to, then yes."

"You must want it less than I do, then." Nan couldn't conceal her disappointment.

Thomas shook his head disbelievingly. "You really are uncompromising! I respect that immensely. I wouldn't have thought it possible to find someone who surpasses me in that regard. But time has two faces—it isn't just brief, it's also long. I would argue its length is trickier to manage than its brevity. Perhaps the real brilliance is the ability to pace things and balance them over time."

Nan was silent for a space. She met Thomas's eyes and saw in them a calm that was absent of the usual sparkle. Presently, she said, "So you could live without me more easily than I could live without you. Is that it?"

"That isn't a matter to compete over," said Thomas gravely. "If it came down to it, should we slow things down a bit, it would probably require more restraint on my part."

"But your restaurant is opening soon. Won't you need my support in that?"

"Only as much as you can give without spreading yourself too thin."

"This is about my bipolarity, isn't it?"

"What do you mean?"

"You're afraid I'm going to have another manic episode and you don't want to be responsible for that."

He gazed at her with a question in his eyes. "I don't want to cause you pain, no. But I'm also afraid it might make you go manic if we slowed things down a bit."

"How can you suddenly want to play it safe? You who have advocated for danger from the very beginning. Can it be that I'm less cautious than you are after all?" Nan's eyes flashed.

"In this area, maybe," Thomas confessed. "I see what my sister is going through, and I wouldn't wish that on anyone."

"So I'm to be punished for my frankness and self-knowledge." She was overcome with sudden sadness.

His eyes softened as he reached over and put his arm tenderly around her waist. "Your frankness and self-knowledge are part of what I love about you."

"If you trust me to know myself, can't you trust that I would be the best judge of my own symptoms?"

"I can, yes," he conceded. "I only want to help make the going smooth for you in the upcoming months."

"So you're not making this decision unilaterally?" she asked, regaining a glimmer of hope.

"I begin to wish I had never brought it up," he said earnestly.

"I could live without you, but I wouldn't wish to."

"Nor would I."

"You know, your presence is more likely to prevent my going manic than it is to cause it."

"For both of our sakes, I'm glad."

"So we needn't slow things down?"

"No." Thomas turned her face to his and gently kissed her.

"Hello, you two," said an upbeat voice behind them. Sue had joined them on the deck. "I've come to find you and bring you down to our jam session. We're hoping we can persuade you to play, Nan."

In the midst of her immense relief at Thomas's relenting, Nan felt gratified to have been singled out to perform. "Of course."

The two of them followed Sue down the two flights of stairs to the lower level, where everyone was gathered. A man they hadn't met yet stood at the double bass ready to play, and another man sat tuning the guitar. Gareth was at the drums, and a long-haired woman in her early sixties, who Nan guessed might be Alice, stood holding the flute.

Alice stepped forward, introducing herself to Nan. "Would you like to head up our little band at the piano?"

"I'll do my best." Nan blushed as she took a seat on the bench. "What are we playing?"

"I'll leave it up to you to decide," said Alice.

Nan suggested a blues song in the key of G. She was a little nervous, since improvisation was decidedly not her forte; memorization and sight-reading were her strengths. But she had enough exposure to the blues and other early jazz forms that she could draw on what she'd heard to hold her own in the musicians' conversation. Nan always found that the hardest part of playing impromptu was to relax into the feeling of the moment—to let the music carry her where it would without worrying about sounding polished or precise. She found that having had three glasses of wine helped tonight in ceding control of the sound. The group followed her as she led them at first to predictable places and then gradually opened up to the sense of freedom of the occasion. She began to respond to Alice and the others in dialogue, and even pushed them to respond to dialogue with her. When they'd finished their piece, the entire room burst into applause, and Alice beamed radiantly in her direction.

"Those were some original harmonies from you, Nan. Thank you. Would you like to lead another?"

Nan thought for a moment, and came up with a boogie-woogie number that she'd played at the retirement home at Ethan's request. She suggested it, and Alice nodded approvingly.

They played a few more such pieces, and Nan's confidence grew with each one. Spirits in the room were high as the group eventually dwindled to the guitarist and bassist, who continued to dialogue with one another in a desultory way after the rest of the musicians had knocked off.

Thomas handed Nan a fresh glass of wine. "That's the most seductive thing you've done yet."

"If I'd known that, I would've played for you a long time ago." As on the few other occasions when Nan had seen Thomas directly after composing or holding her Meetup, her spirits had reached an apex.

"I'll have to put an upright piano in my bar." His admiration seemed to grow in proportion to her confidence.

"I may never leave your bar then."

"So much the better."

Sue came over leading a bespectacled older gentleman with a shiny bald head and a hesitant smile. "Thomas, you remember Evan Andriessen, from Two Shakes of a Lamb's Tail on Gough, near your old bar."

The man shook hands with Thomas and nodded. "I remember *you*, all right. Andy couldn't speak highly enough of you when you worked for him. Where did you disappear to?"

Thomas smiled at Evan. "I have my own bar, in Uptown Oakland now. Soon to be renamed The Chaser." He winked at Nan.

Sue turned to Nan. "Manuel and I met Thomas when he worked at The Esmeralda, and we were the ones who convinced him to move to the East Bay and work for Harry."

"How did you know Harry?" asked Nan.

Sue laughed. "He was one of my most avid extension school students back then. He took every last one of my classes in film studies. Now, in case you're wondering how the owner of an old dive bar in Oakland happened to be a film studies aficionado, he claimed that he fell into it accidentally because the course schedule lined up with his off-hours at the bar. But he had a genuine knack for dissecting films in an impartial way. And he'd seen more mid-century European films than any student I'd taught. Anyway, I'm proud to say that my instinct was right in bringing Thomas to him. He saw Thomas as a son and took him under his wing right away—am I right, Thomas? Harry was always impressed that you spoke so many languages. He once told me, 'In case I get the United Nations coming into my bar some night, I've got a bartender to translate for most of the countries.' Well as a matter of fact, one night, Manuel and I dropped by Harry's soon after Thomas had started working there." Sue's eyes danced as she recalled the scene. "Thomas had been conversing with a handsome Italian man and a Frenchman and his girlfriend at the bar. They were making jokes, and Thomas was rapidly translating from French to Italian and back so the English-speakers followed the exchange. At one point in their conversation—correct me if I'm misremembering this, Thomas—the Italian complimented the Frenchman on his shoes. The Frenchman, who'd had quite a few drinks, thought the Italian was admiring his girlfriend's legs and he said something to the effect of 'Go fuck yourself'—'Allez vous faire foutre.' When the Italian looked at Thomas for a translation, Thomas thought quickly and said, 'Non me ne frega un cazzo'—'I don't give a fuck.' The Italian tossed back the rest of his whiskey, shook his head in puzzlement, and departed without another word. He left thinking the Frenchman was merely dismissing his compliment. In this way, Thomas probably averted a fight." Sue smiled delightedly and continued. "That kind of thing was always happening at Harry's—it was a surprisingly cosmopolitan watering hole for such a dive. I think all the businesspeople and convention-goers who stayed at the downtown Marriott got wind that if they just traveled North a few more blocks to Harry's they could get better liquor at a better price—and better company."

"And was the bartending as interesting as at the Esmeralda?" Nan asked Thomas.

Thomas finished his mezcal. "Well, in one sense it wasn't nearly as interesting."

Sue's laugh rippled tunefully. "He means that he saw a lot more LGBTQ folks at the Esmeralda. Uptown Oakland isn't quite so spicy as Hayes Valley. Now, maybe with your new iteration of the bar . . ." Sue smiled vivaciously at Thomas and then turned back to Nan. "In all honesty, Nan, you've got the catch of the East Bay on your arm tonight.

In all our times frequenting Harry's I can't remember once that I haven't seen women coming on to Thomas, some more subtly than others—many of them drop-dead gorgeous. But you needn't be jealous, because this man has the continence of Scipio. I've never seen or heard of him making one inappropriate move as a bar owner, although God knows he could have his pick of women. We actually once saw a catfight over Thomas—can I tell this story, Thomas?—probably about five or six years ago, when we stopped by on a Sunday night. A raven-haired, emerald-eyed Italian-American beauty with strappy high heels and a green slinky dress faced off against a blond Amazon in a red halter top and white slacks, and the sparks flew. We only gathered what was happening from some gentlemen who had front-row seats when the altercation began. Apparently both had been competing for Thomas's attention for awhile at the bar. The *bellezza* claimed the Amazon had taken the drink she'd ordered, and the Amazon said only sad women had to order drinks for themselves. The *bellezza* replied that she didn't need an ugly guy like the one the Amazon was with, because the hot bar owner had made the drink especially for her. The Amazon retorted that for a face like the *bellezza*'s it was extremely unlikely the bar owner would have made anything short of poison to put her out of her misery. The *bellezza* said, sure, that must be why the Amazon was stealing her drink, and she dared her to test the drink, if she really thought it was poison. Naturally, in response to that the Amazon threw the drink in the *bellezza*'s face. Then all hell broke loose. They clawed, scratched, pushed each other into the bar, rolled on the floor—it was sheer lunacy, and utterly unforgettable. We arrived at the height of the hullaballoo, and while no one else was doing anything, Thomas coolly came over and dumped a bucket of ice-cold water on the two beauties and that stopped their brawling in a second." Sue's voice assumed a touch of regret. "Then you got Benny. That must've been about four years ago? Since then, things have calmed down somewhat on the catfight front. But still, the things we've seen at your bar! And now to think you're becoming The Chaser in a month's time. Manuel and I will definitely be there for opening night."

"That was quick thinking," said Evan to Thomas admiringly. "I'll be sure to visit your restaurant come the new year."

"Oh!" Sue took hold of Thomas's arm. "What if you put some of these stories up on your new website? I guarantee that would attract hordes of people."

"I wouldn't want to be guilty of false advertising," said Thomas drily. "I'm not sure The Chaser can live up to the reputation of Harry's or The Tippler's Haunt pre-Benny."

"Well, I understand if you want to make it under your own steam." Sue sounded disappointed. "But I still say someone should write up all these stories in a book."

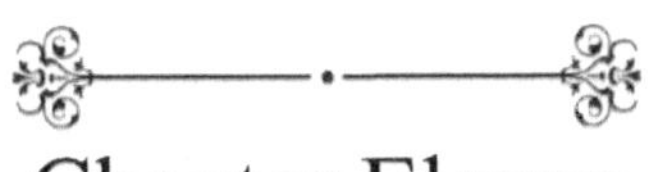

Chapter Eleven

In the Uber home, Nan rested her head on Thomas's chest.

"I wish I could come to your place," she murmured drowsily. "Your place is so much nicer than mine."

"If you come to my place, there's no turning back."

Nan reached up and traced his lips lightly with her fingers. "I could take a bath."

"And then what?" Thomas held her fingertips and kissed them, one by one.

"Then I could walk about naked through your living room as you watch from the sofa. I would lie down on the floor by the fireplace, and you, naked as well, would come and lie behind me, spooning me. You would breathe in my ear, as you do, and whisper tantalizing words while you engulf me with your arms and cover me with your leg. You would make me lose my senses and my reason."

Thomas hovered his hand over Nan's groin for a moment, and then began to trace maddening circles with his forefingers over the area. Even though he did this on the top of her dress, Nan thought she was going to explode from these movements. She reciprocated by slowly running her palm up his hardened length and then resting her hand on it. She placed her other hand on his chest where his open shirt allowed her to access his firm chest. She opened another button in his shirt and began to explore the alluring plateaus and hollows of his torso.

Nan pressed her lips to Thomas's and they kissed in what Nan felt was the most achingly intimate way they had yet. She sensed that Thomas longed for her as she longed for him, that they both equally yearned to ease the tension of their earlier tiff through unprecedented physical closeness. She even surmised that Thomas wished to secure her in a way he never had before—as if their social debut as a couple had activated a reserve of desire and possessiveness in him. As for herself, she was driven wild with thoughts of stunning women throwing themselves at Thomas night after night and him warding them off like a man dedicated to a higher cause.

As they both eventually pulled away, she whispered, "His kisses are like leopards that jump agilely from limb to limb—so controlled, so graceful, so perfectly timed, so soft, so beautiful."

Thomas's laugh was like velvet in her hair. "Ever the evocative poet. Here we are at your place. Would you like to come back to mine?"

As if she'd received an electric shock, Nan's whole body tingled from this exciting proposition. She would have liked nothing more than to join Thomas. It would have fulfilled a number of fantasies she'd had of doing exactly that for the past few weeks. She'd wondered what he might do to her once he had her in his power. How gentle would he be, and how commanding?

Then his words came to her mind from their talk that night: "Perhaps the real brilliance is the ability to pace things and balance them over time." Nan knew from experience that she was good at obsessive brilliance, at compulsive brilliance, at manic brilliance. Could she set to herself the challenge of mastering long-term, steady, evenly-paced brilliance? She determined in that brief moment to attempt it, and to do so by continuing on their path of the long seduction until their act of coming together was of the utmost necessity. Now, more than before, she had the feeling they would both know when that time had come.

She kissed him. "I'd love to. But not tonight. I'll see you tomorrow for our ride?"

He nodded knowingly, as though he'd read her thoughts. "See you tomorrow."

The next day, Nan suggested they ride out "the Flats," as she'd heard a sprinter friend call the at-first rolling and then largely flat route that fringed Crockett, Martinez, Briones, and Pleasant Hill. The Carquinez Scenic Byway was one of Nan's favorite stretches of ride, as the very first time she'd done it she'd seen a family of red foxes at six in the morning and had an engaging conversation with two old men who were walking near the start of it. To their right, the gently rounded hills that only two months before had risen like giant piles of desert sand now wore a luxuriant green coat. Today Nan led the way, since there were a number of variations they could do on this route, but she had a particular one in mind.

Thomas had apparently secured Luca, the Cuban chef recommended by Elian, and they planned to draw up a menu on Monday for the restaurant. Luca proposed to have a tasting for the bar-bites menu before Christmas. Luca and Thomas would vet potential kitchen staff—two more line cooks, two prep cooks, and two dishwashers, to start with—and aim to have them onboarded by the start of the holiday break. In the meantime, Thomas had advertised for wait staff. It seemed his current staff were already instrumental in putting out the word that his bar was hiring. He was interviewing a dozen applicants in the coming week.

When they took a break at the Carquinez Bridge overlook point, Nan showed him pictures on her phone of the various designs she'd made for the bar-restaurant's logo. Thomas particularly liked two of them that mimicked the style of Tour de France memorabilia. Thomas sent her a link to The Chaser's website so that she could read through it and give him input on the overall layout.

He asked after Margot, and Nan briefly told him about Aaron, as well as Margot's plans for the next exhibit at the gallery.

"Do you have a Margot?" she asked, as they descended into Martinez.

"I have a few close friends who are scattered over the country—guys I went to school with, and one guy I spent a lot of time with in Vienna with whom I keep in close touch. I also have a friend named Len out here—he and I used to race and train together before I knew you. He's a real character—full of wild energy, and a mix between a bad-boy cyclist like Peter Sagan, on the weekends, and a responsible mortgage specialist, on weekdays."

"Why did you guys stop riding together?"

"He and his wife had a baby just before you and I met last December. That really slowed him down on the bike."

"You should invite him to ride with us one of these days. I'd love to meet him."

"I will. He's actually been texting these days saying he's keen to get back in shape for racing by doing training rides again."

For the rest of the ride they chatted about the conversation they'd had with Sue and a few other guests after their talk about Harry's the previous night. Since her early days teaching in the UC Berkeley Extension School, Sue had become an associate professor in film studies. Much of the discussion had centered around the creative ways that Hollywood had devised for continuing to make movies and shows during the pandemic. At last they climbed Pinehurst and, slightly out of breath but triumphant, they paused to say goodbye. Thomas was riding down the steep Shepherd Canyon by way of a shortcut to his house. He had a lot to do at the bar that afternoon, and needed to get a head start.

"Thanks for the great ride." He tucked a strand of hair behind her helmet strap.

She nodded as they both leaned in for a kiss. "See you soon, I hope."

"I count on it."

She watched as he swooped elegantly down into the canyon, reciting to herself part of a speech of Imogen that she had memorized from *Cymbeline*: "'I would have broke mine eye-strings, crack'd them, but To look upon him, till the diminution Of space had pointed him sharp as my needle—Nay, followed him till he had melted from The smallness of a gnat to air, and then Have turn'd mine eye and wept.'"

Margot was coming over to Nan's early that evening before she met up with Aaron, and she said she had "a lot to tell." In the interim, Nan shipped off a few more eBay items, performed the gloomy task of balancing her checkbook, and worked a bit on her problematic second movement.

Nan had put a bottle of white wine in the fridge for her and Margot. When Margot arrived, they kissed, and Nan ushered her in, taking her coat and bag.

Margot sighed. "Why does everything always have to be so complicated with guys?"

"Uh-oh." Nan poured them both glasses of wine. "Come take a seat."

They sat on the floor in the living room.

"Yesterday, Aaron and I were talking on the phone and he confessed that he's been feeling overworked and underappreciated at Facebook. He said they put a lot of pressure on each employee to cover a range of responsibilities. He wants to be in a position of having a better work-life balance than he currently does." Margot sipped her wine. "He and I have managed to see each other about twice a week ever since we first met up, and it's been great. But he's always stressed out, and we both end up feeling the time crunch when he's got an upcoming deadline—which is pretty much always."

"But he loves what he does, right?" asked Nan.

"Yeah, no question. He believes in the whole mission of the company. But ultimately, he wonders if he shouldn't move on and try for a product manager position at Google, where the work culture is less demanding."

"That sounds fantastic." Nan put down her glass. "He's ambitious, and yet he wants to make sure he has time for you, right?"

"Yes, in theory," said Margot. "But if he makes this move, it's going to tie him up for at least two to three months while he gets adjusted at Google. He said as much himself."

"I see. So it's a question of a short-term sacrifice for a long-term advantage. Has he already decided what he's going to do?"

"He said he's applying for the Google job, and will let their decision determine what he does next."

"Is it pretty likely, do you think, that he'll get the job?"

Margot smiled. "I think so. He's brilliant. I don't see how Google couldn't want him."

"Well, did he suggest any ways to make things work in the meantime—with your relationship?"

"He said we could still try to meet up once a week, even if just for a few hours."

"Can't you exercise with him, or have lunch with him, or anything small like that, just to be together a little?" Nan suggested.

Margot nodded. "I actually proposed both of those. I don't work out in the gym, but I can go on runs with him. And neither Facebook nor Google is that far from where I work, so I could go eat with him at either place. He seemed to like these ideas. But I'm disappointed that *he* didn't suggest them. It was all me doing the brainstorming. He's so proactive in so many areas—why not in this one?"

"I think you may have put your finger on the problem in one of our talks about Thomas recently," Nan submitted. "If he's stressed out and making a major transition, his mind may be stretched to the breaking point. But he's committed to you in all this, right?"

"I'm pretty sure. And he's definitely worth trying to make it work. I feel as if I could easily fall for him, Nan."

"It sounds as if you both may already be falling for each other." Nan reached over and put her hand on Margot's.

"He's such a good kisser." Margot sighed. "It would be a shame to have to put that on hold."

"You won't. I'm sure he wants to make it happen as much as you."

Margot told her about some of the intimate moments she and Aaron had shared in the last two weeks. They spoke of their respective Thanksgivings, and Nan told her about the party she had gone to with Thomas, mentioning his offer for Nan to come back to his place after the party.

"Oh." Margot shook her head. "I don't think I would've been able to resist that offer."

"I was so close to taking him up on it, Marg. But I begin to respect his vision of our long seduction. I see more of its advantages now, and fewer of its disadvantages. The main pro is that as we get to know each other better, we seem to get more and more keyed up for the deed—whenever we end up doing it. Another benefit is that I increasingly see the ways in which he inspires not just admiration, but awe. I can't imagine that won't play a major part in our lovemaking. We also get to use words to build the eroticism of being with one another." Nan took a sip of her wine. "But above all, each time we talk, there's this indescribable dissonance that ends up being the sexiest thing I've ever experienced."

Margot reflected. "When you first told me about this long seduction, I'll admit I was worried that he might be the sort of man who, once both of you have sealed the deal, would begin to lose interest or take you for granted. Now it sounds like the complete reverse: he seems incredibly sensitive to the long-term possibilities that this could lead to, as well as the ways this could be really exciting in the act itself."

"*Really* exciting." Nan smiled. "Shifting gears a bit, did you sell all of the artworks in the exhibit?"

"A good percentage of them, yes." Margot beamed. "We'll be taking everything down during the first week of January. Then we have one week before we start putting the new works up."

"How do you do it all so fast?"

"Timing is everything—we'll be spending all of December arranging with our artists for transport and installation. The schedule is really tight and precise, and our success depends on several critical people."

"Whom will you feature in this next exhibit?"

Margot told her about some of the Taiwanese and Taiwanese-American artists and their pieces. She described the write-ups she'd been working on for the program and the walls of the gallery. These blurbs would help visitors understand the history and socio-cultural context for the works. "I've been gathering the oral histories of older Bay Area residents so that I can weave them into the write-ups and run video clips of them sharing their stories in one of the upper rooms."

"How do you define 'Taiwanese'"?

"Anyone with roots on the island, and who identifies as Taiwanese."

"This is going to be *amazing*, Marg."

Margot's smile widened. "And my parents will be coming out at the end of January."

"Will I get to meet them?"

"They want you and me to give them a tour of the city. I said, 'Which part?' They said, 'Fisherman's Wharf and UC Berkeley, the Presidio and the De Young.'"

"That'll only take a few hours." Nan laughed.

They chatted about their upcoming holiday plans, and then Margot stood. "I should be off."

"I'll walk you to the BART."

Margot was headed to Macarthur station, where Aaron was meeting her to take her to dinner. When they reached Rockridge station, they hugged tightly, and Nan said, "Next time, I'll come into the City and we can take your crazy hill walk." Margot had a series of hills she hiked that made even Nan's heart rate soar.

"Okay. Good luck with the composing tonight!"

Nan watched her go and was flooded with love and gratitude for such a good friend.

♪ ♪ ♪ ♪

Thomas had texted her by the time she got back home. "We're on for next Saturday with Len and another rider from the group named Carl."

"Cool. Where are we headed?"

"Three Bears to Pig Farm and Reliez Valley," he texted.

"How's the bar tonight?"

"Hopping. It's always like this once the holidays get going."

"I'm glad to hear it."

"How's the composing?"

"Just starting it now, after I scare up some dinner."

"Talk to you in a bit then."

Nan immersed herself in writing until around 11, when she decided she'd better try to get to bed at a decent hour for once this week. As she went to lock her door for the night, she noticed groups of ants trooping up and down the cracks in the plaster on the wall. She followed their path and saw that they were crossing the counter, traveling up and over the refrigerator, and down to the trash bin on the other side. She'd had ant infestations on several occasions before this, especially when the weather was about to turn extremely hot or cold, and she kept Terro ant bait containers on hand for just such times. She slit open a container, placed it on the counter, and got ready for bed.

The outside-facing wall on which the ants crawled upstairs was the same wall against which the head of her bed was set downstairs. As she wrote in her journal, she noticed a few ant scouts going up and down the moulding of the door by her bed, and even a few ants exploring along her sheets. She crushed them and moved her bed out from the wall by a few inches.

The next morning, the ants' activity both upstairs and downstairs had increased fourfold. Hundreds of them were all over the wall, and, disconcertingly enough, they seemed to have largely ignored the bait she'd set out for them. She placed out another panel of bait, and called Thomas.

"Good morning." He yawned a bit.

"I hope I didn't get you up." It was just after nine.

"I'm glad you did," he said. "I can't think of a more pleasant way to start the day than hearing your voice."

"It's nice to hear yours too. It sounds deeper this early. How late were you at the bar last night?"

"Till about 11:30. All the regulars were there in force. And we had a marriage proposal."

"Oh, how romantic! How did it play out?"

"He said he had tickets to the 49ers game tonight, and passed her the envelope. She felt something unusual in it." She could hear the smile in Thomas's voice. "She looked totally shocked when she opened it up. I think she really had no clue this was coming."

"But shocked in a good way?"

"Yeah, her expression went from surprise to over-the-moon in half a second. Thankfully for everyone, he didn't opt to go down on one knee or do any gushing. He just said some quiet, sensible words like, 'I love you. Will you be my wife?'"

Nan laughed. "It's good to know what you consider reasonable in a proposal."

"Were I a woman, I don't think I'd be flattered to have my partner abase himself or get sensationalist, however public or private the proposal."

"I happen to agree with you," said Nan. "But I don't think we're typical. Much as I prefer this subject to what I'm about to segue to . . . have you ever had ant infestations?"

"Often. I have a pretty good system for dealing with them."

"Please tell."

"After putting out the ant bait containers, I sprinkle diatomaceous earth around the edge of the area where the containers are. This corrals the ants in and cuts up the ones that try to pass over the earth."

"What do you do when they ignore the bait?"

"I haven't had that happen yet."

"If they don't go for the bait over the course of today, I may ask my landlord to have an exterminator come and spray the wall. They're crawling all over the area of my bed downstairs. I woke up last night with a few on my arm."

"That's no good. Yeah, I've never found that the natural citrus sprays they sell at the hardware store work. And their smell is pretty overwhelming."

"What are you up to for the rest of the midday?"

"Catching up on news, reading some French, and getting in a workout. I'd offer to take you to brunch, but I know you have your first lesson at 11."

"Yeah, Chloe's a sweetheart—such a wonderful first student to have for the day."

"Let me know how the ants turn out."

They signed off, and Nan geared up for the long slog that day. She had eggs and toast while reading "The Morning," a collection of short highlights from *The New York Times*. She then examined her December calendar. Her trio's concert was this next Saturday evening. The following two Saturday afternoons were the recitals for her various students, one at a house around the corner, and the other at a house in Upper Rockridge. She'd been able to divvy up the students so that seven of them of roughly the same ability and level were playing at each recital. This had taken some dexterity, and Nan was pleased with the results of her efforts.

Lessons went well that day. Most of Nan's students were clearly excited by the prospect of performing at their recitals, and this spurred them to focus even more intently on their work. By the evening, however, when Nan started to make dinner, the ants had gone from bad to worse, and it was undoubtedly the same colony infesting her kitchen and bedroom. They had hardly touched any of the three panels of bait that she'd put out. She texted her landlord asking him if he might have an exterminator come over. A half-hour later, he texted to say that he and his family were out of town, but he could arrange for an exterminator to come later in the week.

The idea seemed bearable for the hours that she worked upstairs. But when bedtime came around and she saw how many ants were crawling around up and down the wall, on the nightstand, and in her bed, she began to despair. At this point, not only did she have no more space to move the bed out any further, but doing so seemed futile, given the sheer number of ants in all directions.

Once more, she called Thomas.

"What's the status update on the ants?" he asked.

"They're all over the blankets, pillow, and sheets," she said anxiously. "My landlord can't send out the exterminator until sometime this week. I honestly don't think I'm going to get a wink of sleep tonight."

"Yeah, you can't exactly put bait or diatomaceous earth in your bed." Thomas was silent for a moment. "You could come over and sleep in my bed. I can sleep on the couch."

Nan was moved. "That's really generous of you, but I can't put you out of your bed. I'd be glad to sleep on your couch, which is still much better than a bed full of ants."

"Let me come pick you up. I'm just leaving the bar."

Nan was decidedly nervous as she took her various pills and gathered together a few night things to take to Thomas's place. Was this a mistake? Would neither of them sleep as a result of this decision? Somehow the risk of not sleeping at Thomas's seemed infinitely preferable to the guarantee of not sleeping at her own. Still, to calm herself down, she hummed some of the motifs from her quintet. She packed her staffed notepad and pen, in case some ideas should come to her in the next twelve hours or so. Her heart was pounding in her chest as she got into the passenger's seat of Thomas's car and he leaned in to kiss her.

"You're nervous," he observed. "Has the situation in your apartment got you 'antsy,' or are you so afraid of what will happen once we reach my place?"

His words made Nan's heart leap into her throat, and she couldn't speak immediately. "I . . . I hadn't gotten the chance to really think it through yet."

"We can treat this as another opportunity, you know."

"Oh . . . " she stammered, "you mean, for . . . "

"For getting to know each other."

"Okay. I see."

Thomas smiled. "Take a few deep breaths." He put the radio on, and tuned it to the classical station. "Does this help?"

Nan nodded. "A little."

When they entered his house, Thomas flicked on the lights to the living room, and watched as she removed her jacket and placed her bag on the chair. "Would you like to take a bath?" His voice had provocation in it, and his eyes glittered.

"You're making this very hard for me." Nan reddened. "How can you be so calm?"

"It comes naturally." He took a step towards her, allowing his eyes slowly to sweep her body from her eyes down to her thighs and then back up again.

"When you do that, I feel as if I might as well be naked."

"That's the idea," he replied with maddening steadiness.

She blushed afresh, longing for him to break the tension with some kind of movement, but instead he held her fast with his gaze alone. She swallowed and found she couldn't look away. His eyes had a kind of summons in them—as though daring her to play at something new.

"Don't you think it's time we explored some of the possibilities of our different characters?" he asked.

"Haven't we been doing that all along?"

"Not to the extent that we might now."

"Do you mean explore our preference . . . for a certain dynamic between us?"

"Precisely."

Nan watched, riveted, as he took a seat in the armchair, crossing one ankle over the knee of the other leg in an easy way. Then he opened the drawer in a little table beside the armchair, took out a box of matches, opened it, and flipped it upside down over the hardwood floor. Matches flew everywhere on the otherwise pristine floor.

Startled, she moved to pick them up.

"Leave them." A faint smile played about his lips.

"But . . . why?"

"Just an experiment. Come a little closer."

She took a tentative step towards him.

"Do you trust me?"

"Sometimes," she said honestly.

"What about now?"

"I do, and I don't."

"What are you most afraid of losing?"

"My freedom."

"Can you trust me with that for tonight?"

Nan hesitated. "I'm afraid . . . but, I think so, yes."

Thomas held her gaze. "Remember, this is only our first exploration into the unknown."

She nodded.

"It's an added bonus to your being a woman of the mind and a lover of contention—the fact that you also prefer what you most fear," said Thomas. Then he added thoughtfully, "It suits me perfectly."

He rolled up his sleeves slowly, allowing her to take in the prominent network of veins on his broad forearms. She still stood expectantly, mesmerized by his every move.

"What did you bring to sleep in?"

"My top and my pajama shorts."

"What are they made of?"

"The top is cotton, and the shorts are silk."

"Go into the bedroom and change into them. I'll wait here."

Nan heard a quiet but firm note in Thomas's voice that matched his authoritative profile as she had admired it in the train the other day. She nodded amenably, and passed with her bag to the back bedroom. As she slowly removed her clothes and put on her top and pajama shorts, she felt as if she was being swept up by a force beyond her understanding. For once, she was incapable of watching herself from an outside standpoint or of considering too much what her next action would be. The fact that she was changing in Thomas's bedroom, surrounded by all of his things, heightened the sense that he had full command of the situation. After she had changed, she noticed a pile of clothes at the edge of the room that looked like Thomas's workout clothes. She couldn't resist: she took up one of his shirts and breathed in the heady, masculine scent of his sweat.

Emerging from the bedroom, she felt giddy with the possibilities of what might happen next. She approached Thomas, who, true to his word, had remained seated in the armchair. As she passed him, he brushed her wrist with his fingers, as if considering seizing it. This subtle motion created havoc in Nan's mind, suggesting as it did his formidable power.

She turned and stood once more, not two feet from where he sat, facing him.

"Go into the kitchen, fill a tall glass with water, and bring it here."

She didn't waver in carrying out his command. When she returned and offered the glass to him, he took it and, to her horror, calmly poured out a fourth of it onto the floor. She inhaled sharply and reached for the glass to stop him from pouring out any more. He held it just beyond her grasp and said, in a peremptory tone, "Since you tried to interfere, you will have to pour out the rest of it yourself."

Nan shook her head vehemently. "I can't."

Thomas rooted her to the spot with a resolute eye, held out the glass to her, and declared with quiet authority, "You will."

She closed her eyes, wrestling within herself. Having been raised Quaker, she had never cared a fig for what her peers had done or said—peer pressure had no hold on her, and she had always been the odd person out in school. But she had never minded this—had always, in fact, been proud of her internal and external strangeness. Yet here it was not a question of doing something merely because other people were pressuring her to do it. This moment was pregnant with challenge, danger, competition, a meeting of the wills.

On the one hand, the idea of turning the glass that Thomas offered her upside down and spilling its contents revolted her to the core. If there was anything Nan hated, it was incontinence: she couldn't stand a leak anywhere, whether in a faucet, a roof, or a container. It violated all her deepest principles of control and discipline. Seconding her instinct for containment was her native rebelliousness, which drove her to question every command she was ever given. It had been one thing to obey the small, reasonable orders Thomas had given her; it was another thing entirely to follow his current, arbitrary dictate. Lastly, Nan's strong competitive streak urged her to test the limits of his will. If she refused, what would he do?

On the other hand, when she'd looked into Thomas's eyes, she'd seen a steely determination that frightened and thrilled her at once. His presence had become almost demonic. In that moment, she had a presentiment of how their standoff would inevitably end: his sharpness of purpose would cut like flint through her own wavering resolve, and he would get the better of her. Like his masterly moves on the chess board the other night, his far-seeing tactics in seduction would outmaneuver her most creative dodges. Competitive as she was, she found herself aching for her opponent, a condition that precluded opposing her will to his. Meanwhile, her compulsion to be in thrall to him shunted aside all consideration of the rightness of things—or, rather, made what was once wrong seem right. Remembering how he had adjured her to trust him with her freedom, she shook from head to toe at the thought of what she was about to do, an action that went against all her deepest instincts. She opened her eyes, took the glass he held out to her, and slowly emptied it onto the floor.

To her great surprise, in response, his eyes flared up with fire and longing. "I never knew how strong you were before this moment. Sit here," he said, removing his arm from one of the chair's armrests.

Now, perversely, at the time when she most wanted to do what he'd commanded, she also strained to resist his order. She saw how he exulted in her submission when it was most painful for her to submit, and she shrank from the idea of making the way so easy for him. She shook her head and took a step back.

He regarded her darkly with his eyelids partially lowered, saying in a husky, hypnotic voice, "Your restraint deserves reward. I won't make further trial of you tonight."

When she continued to hesitate, he uncrossed his leg and leaned imperceptibly towards her, lust clouding his eyes. "You are too beautiful when you defy me. It whets my appetite for you. But you can't imagine I will let you escape now, when you're completely in my power."

She trembled from the force of his words, even as she felt exhilaration at having his darkened eyes and deep voice directed towards subjugating her. In that moment, several scenarios played out in her mind, and they all involved the unnerving forearms he had exposed to her, which looked fully ready to secure her should she choose not to yield to him. She recalled these forearms pulling on the handlebars of his bike when she had sprinted against him on flats. Testing these arms now by continuing to hold out against his redoubtable will would only compound her longing to end up consumed in those arms and thereby ensure his conquest. Yet, she still refused to concede.

He smiled, as if reading her thoughts. "I believe we are of one mind. We both enjoy the taste of my power over you. And we both know how this will end." He leaned back comfortably in the chair and allowed his eyes slowly to rove over her body. "I delight in your disobedience, for it augments my own achievement in mastering you."

His assurance made her vibrate anew and her heart began to beat at an accelerated pace as she waited for his next move. She was drenched with desire, even as the suspense had tightened her nerves to the breaking point.

He slowly began to undo the clasp of his leather belt, and the attention he directed to this task made her concentrate upon his movements. Was he going to take the belt off? What would he use it for? As she was puzzling over these questions, he took advantage of her distractedness and startled her by swiftly seizing her hand and pulling her forward; in one smooth motion, he wrapped his other arm around her waist and drew her down to his lap so that she was seated sideways over his thighs, her back against his chest and her backside over his groin. Placing one of his hands on her thigh and the other on her breast, he pulled her in close. The entire action took no more than three seconds, and, stunned as she was, Nan was forced to admire the deftness and elegance of his maneuver, which resembled a dancer's pulling in of his partner after swinging her out.

"Are you comfortable?" he whispered in her ear.

Her torso and limbs quivered at his touch, but she managed to nod. Gone was all thought of struggling, now that his body submerged her on all sides like the resisting element she swam through in her dreams. Gone was her desire to test him further, once she felt the unmistakable benefits of yielding. She closed her eyes and moaned slightly, giving in to the profound darkness that encompassed her.

"I could ravish you just like this," he growled, moving his hand up to her neck and lightly pressing his fingers and thumb to her carotid arteries. She shuddered again, finding herself caught between his hand and his lips, and ceding complete control. "You're so trusting, so pliant . . . so utterly mine."

She was melting between her thighs, and felt the rise of him that told of his own urgent desire. She sighed falteringly and dug her hands into his quad muscles, tilting her head towards his with a pleading hunger. He answered by devouring her lips and tongue with a greediness that surpassed even the pressing movements of his hands as they roamed raveningly over her chest, belly, and groin. His conquest in this moment was complete, and she found herself instinctively swiveling on his axis, gyrating about the rigid heat at the center of his body. He groaned, allowing her to sink her hips more deeply into him.

In between kisses, she said in a broken voice, "I can't bear to stop. Please."

"We won't," he declared hoarsely. "Come this way."

He led her to the bedroom, where only the light from the hallway illuminated the bed at the center. He stood opposite her in front of the bed and directed her to take off her shirt. She did so unwaveringly. His eyes seemed to feed on the sight of her bare breasts.

"Now take off your shorts."

Once more, unquestioningly she did as he instructed. He breathed deeply. "Turn around. Slowly."

She did so. When she had made a full turn, he murmured, "You're even more enchanting than I'd imagined."

She stepped closer to him, searching his eyes for permission and reading in them assent, before she began to undo the buttons of his shirt. As she reached the last one, he hastily pulled it off. Now it was her turn to inhale sharply. He was even more of a stunner than she'd anticipated.

Now with increasing eagerness she undid his belt, unzipped his pants, and cast them down. Then, looking into his eyes for final approval, she slid down his boxer briefs. She stood back and shook her head at what she saw—lithe, muscular limbs that looked ready to spring into action, strong shoulders, chiseled abs, and arousal that spoke of what must come next.

He said, "I have any number of ideas about how this next part might play out."

"Can we do the simplest of them tonight?" she said with growing insistence.

He laughed softly and took a step towards her. "Remember, this is a work of beauty, a piece of art."

She knelt before him and took his length in her hand. "I find this a work of beauty and a piece of art." Then she could speak no more, and nor could he, for what she did. He moaned as she moved her lips over him and took him in her mouth. After a while, he gently pulled her up and eased her onto the bed. He lowered himself over her, guided her fingers to the top of her groin, and hovered above her a moment—as he had done in her dream—with hungry, possessive eyes. She arched her hips into him, and as he filled her, they both momentarily lost their breath. She clutched at his back with one hand as he moved in and out steadily and watched the fingers of her other hand trace circles over her clit. She broke first and let out a long wail. As she came a second and third time, he too groaned deeply, his head wrenching to the side with the intensity of his climax.

As he lay beside her afterwards, moving his hands softly over her breasts and belly, she said, "I didn't mean for this to happen tonight."

"You're not sorry it did?"

"Not now. It was as you said it would be—there was a kind of necessity to it that demanded we take it all the way." Nan reserved her thoughts of how difficult she was finding it to resolve the conflicting images of Thomas in her mind, where he appeared by turns sympathetic, safe, and secure, and dominating, dangerous, and demonic. She relished both sides of him, but marvelled at their contrast—as in the seventeenth-century Dutch paintings her mom had loved where brilliant rays of sun glinted through stormy grey seascapes.

"I have the ants to thank for this," he said.

"Them, and our frailty," she mused. "The greater our resolve became, the more our consummation mocked us in the end."

"This isn't the end," he countered. "It's only the beginning."

♪ ♪ ♪ ♪

In the middle of the night, she woke to find his hardness against her from behind, his hands caressing her curves. The two of them began to move in concert as he ran his fingers up and down the front of her body as a cellist plays his instrument. When she tipped her lips up to his, his tongue by turns caressed, teased, and challenged hers, keeping her ever guessing as to its next move. He took her from this spooning position and their ensuing cries, though softer, expressed no less intensity of release.

The early morning saw a reprise of this as he laid playful kisses on her collarbone, neck, and cheek. Before she knew it, he had pinned her arms above her head and, smiling down at her, moved so that he was directly above her. Intuiting her readiness from the way she shifted her hips in eager anticipation of his entry, he took his time, teasing her with his proximity, a tip, a bit of head, and then none. She wanted to reach out and grab onto some part of him or help bring him into her, but his hand pressed her own hands firmly into the pillow. He drove her insane in this way for what seemed like an eternity. Only after she had begged him several times, promising him she would do anything he wanted, did he relent and slowly give the entire rigid part of him, using his free hand to touch her as he'd seen her touch herself the night before. She gasped as he did so, her pleasure quickly mounting as she felt his own intensify. They reached a peak together, their groans merging in unison before dissolving into separate strains of whimper and sigh.

She watched the morning light that came through the skylight transform the room's tint from violet to blue to pale yellow, and then to a rich gold as they lay in bed past nine. Her head rested on his chest, and she drew patterns on his midriff, while he ran his fingers through her hair idly.

"Are you as ravenous as I am?" she asked at last.

"Yes, I've just been taking mental stock of what I have in the fridge. I think I can rustle up something."

"I'll help you."

But they lingered a few more minutes, as if both under the spell of the morning's calm beauty. Finally, she kissed him and made as if to get up. At the last second, he hooked his arm around her waist and swung her back down again, evoking a ripple of laughter. "Not so fast," he said. "There's a round four, you know."

"There is?"

By way of reply he kissed her and pulled her on top of him. Instinctively she began to rock back and forth as he admired her from below, gripping her buttocks in his hands. This time was longer and steadier, but its finale was no less full of rapture. She collapsed on his chest, where the rapid beats of his heart testified to the forcefulness of his own unloosing.

"I'm afraid we're both addicts already." She reveled in the smell of his neck and the rough feel of stubble on his jaw.

"It's a healthy addiction," he said in his deep morning voice, pulling her even closer to him. "Your skin is so smooth and soft. I don't want to stop touching it."

"I've been thinking about the last nine hours," she said. "And wondering if the rarity of them lies in the fact that we've been leading our bodies on for so long—or if our lovemaking is the fruit of the mind. When I spoke to Margot about this the other day, I would have thought the latter. And I know that my own high-strung and obsessive cast of mind tends to create scenarios like last night—and like many of our encounters over the last few weeks. But then I come back to the question of the truth of things, and feel the forcible realness of all I've lived through, and I become conflicted again."

"I would say our lovemaking is a result of the mind as it works in concert with—and in opposition to—the body," he said thoughtfully. "I think your conflictedness is precisely why this sort of seduction lends itself so well to contest—contest within yourself and contest with the other person."

"Do you mean that each time we make love, we're saying to ourselves, 'This is and isn't the truth about myself and the person I'm making love to'?"

"Yes, as far as the inward contest goes." He seemed to be working much of this out as he spoke. "That contest makes up the doubt of the seduction. The outward contest—the discord with the other person—is a battle of discipline and restraint, as you've seen. This contest involves two strong wills encountering one another intellectually, emotionally, and physically—often all at once. We can't say that the seduction relies merely on the outcome of the clash—though for your and my particular purposes the clash may end with my getting the better of you—but, rather, on the critical experience of having confronted one another, sharpened our wits and skills against one another, and tested one another."

"But isn't the more vulnerable person liable to suffer in this battle of discipline and restraint?" she asked.

"Not if we go slowly enough to allow ourselves to trust each other. That's where the 'long' part comes in." He ran his fingers down her arm. "We have to make sure at all times that we're both more or less at the same point of emotional commitment."

"That still seems very dangerous," she reflected. "Because to me the long seduction borders so closely on a game—although you insist it's a form of art. If it *is* a competition, someone's heart may break."

"We need to make sure that doesn't happen. I agree with what you said a while back: good communication is key."

"What would've happened if I'd said no last night to tipping out the water?"

"I would have brought you around eventually," he said with assurance. "But you put up an impressive struggle."

"It seems as if you're doing all the testing," she mused. "I don't think I've tested you very much."

"Weren't you testing me when you told me about your bipolarity?"

She was puzzled. "How so?"

"I don't think you were making a cry for help. I think you were testing my staying power and trying to see if this would push me away."

She realized he was right. "It's very likely. I have a perversity in me that seeks to shake things up when they seem too stable."

He chuckled. "You seek discord out of perversity; I seek it by nature."

She was just drifting off to sleep again when she became aware that he was shifting her and getting up. She lay drowsily on the pillow, half asleep and half awake, as she heard him moving about the kitchen. She thought of the trust she'd shown Thomas last night. That trial reminded her of the trust games they used to play in drama groups when she was in high school—passing an upright person around the circle; walking in pairs keeping eyes closed and using only one leg; falling backwards on a partner. Those games had always been cathartic for Nan, as they provided a formal framework for ceding complete control. Much like the freedom she'd felt improvising with the musicians at Sue and Manuel's, the weightlessness that came from trusting another person to that degree felt like losing oneself. It occurred to her that the step from yielding control to living in the present moment was a short and feasible one.

Meanwhile, the rich smell of bacon, fried eggs, and toasted buttered English muffins wafted throughout the house, and her stomach rumbled expectantly. She got up, brushed her teeth, and went to join Thomas at the stove.

"Can I help?"

"You can take down some plates and mugs from that cabinet. I've made us French press coffee. If we weren't so hungry, I'd have taken the time to make us pourovers. But I think you and I need food sooner rather than later."

She sighed with pleasure. "You could be frying up a leather boot and I'd wolf it down at this point."

"Nice to know you're not picky."

They brought everything over to the bar counter and sat next to each other, their legs and shoulders touching. She laughed, recalling something. "You know those body-language experts who are always doing analyses of celebrities' relationship statuses in tabloid magazines?"

He smiled. "I can't say I've read many of those, but you've given me a pretty good idea of them in that brief description."

"I wonder if one of those experts would guess that you and I had slept together for the first time last night, based on looking at us here now."

He considered this. "I would guess that only the fact that it's breakfast we're eating together would give away our having spent the night together. You and I have been pretty close physically since our first date. This could be any of our dates in the last few weeks."

"You don't think we both have a special glow to us?"

"Picking up on that may be where the expertise comes in." He speared a slice of bacon, tucking it between his egg and English muffin. "I liked what that woman Claudia said the other night: the French cherish the kind of love that consists in doing small things together."

"To me, these small things never get old. At least, not when you're doing them with someone you care for," she agreed.

"What've you got planned for the day?"

"I don't have to be back to teach at my place until three."

"Stay here until then and work, if you'd like. I usually take Mondays off from the bar, but this evening I'm interviewing applicants for the server positions. I should be back by seven or so. I could make us dinner and you could stay over again . . ."

"That's right—I'd forgotten that the ants probably won't have gone away yet." She thought for a moment. "I don't want you to feel you have to host me until the exterminator comes. I can always go sleep at Bess's place."

He took a sip of coffee. "Don't be silly. Do I look as though any of this is a burden? To me, it feels like a bit of much-needed vacation."

She nodded. "Last night and this morning felt like that to me as well. But what if you grow tired of me?"

"I'll be sure to chuck you out as soon as that happens."

After they'd cleared everything away and put the dishes in the dishwasher, she took a seat at the dining room table to work on her second movement while nursing the rest of her coffee. Thomas went into his office to work on the social media advertising for his restaurant. After about an hour, she took a break from composing and opened up his website on her phone. As she read through it, she jotted down notes for possible changes; she was struck, however, by the utter absence of typos or grammatical errors. Clearly Thomas was extremely meticulous with his writing. When he emerged from his office an hour after that, he saw that she was on her phone and he came over to her, placing a hand on her shoulder.

"Am I interrupting?" he said.

"No, I was just starting an email to you with suggestions for your website."

"Why don't you bring your ideas in here and we can look at it together on my laptop?"

They walked through the entire website with her notes in hand. She pointed to places where he might incorporate the cycling logo or a similarly themed image or bit of text. He showed her the placeholders for the food menus, and she gave him her input for the drinks menu, which already appeared on the site in its entirety.

He moved an image across the page to the opposite corner. "I'm thinking of having Elian and Luiz prepare a few seasonal cocktails that could pair nicely with the seasonally themed dishes."

She heartily endorsed this. "I love it when restaurants do that. It's like a *prix fixe* menu—it saves you having to make one more decision."

As they stood at his standing desk—which she loved for the way it spared the hamstrings, lower back, and hips—a text came in on her phone. He glanced down at it.

"Who's Dave?" he asked.

She looked down and opened up the text, showing it to Thomas. It read, "I have a few gigs coming up over the next few weekends, if you can make any of them." Dave had listed the dates, times, and venues. At the end, he'd signed off, "Great meeting you."

"He's the son of one of the residents at the retirement home. He plays saxophone. He wanted to give me the times of his group's upcoming gigs in the area. If you'd like to go to one of them . . ."

Thomas nodded. "We can. You'd let me know if he's the sort who wants to sleep with you, right?"

She blushed. "He's innocent enough. He may've seemed interested, but if you and I show up together, he'll get the idea."

Thomas held her eye. "You're a very open, sweet person, Nan. And you've got the genuine Minnesota niceness on top of it all. Men can easily misinterpret your friendliness. I wouldn't have you change it for the world. I'm just alerting you to the possibilities."

She laughed. "Says the hottie bar owner who has women and men drooling over him every night."

"That's just good business," he replied, half sternly half roguishly.

For the rest of the afternoon, they worked side by side at his desk, she writing her composition and he translating a few passages of Lucretius from the Latin. At 2:30, he drove her back to her place on his way to the bar.

"See you around 7:15?" he said.

"See you then."

The ants' takeover of her kitchen and bedroom had gotten exponentially worse since yesterday, and the bait panels remained largely untouched. Nan was relieved to know she didn't have to sleep here tonight. She felt like thoroughly washing her hands of her apartment for awhile. She texted Bess to remind her that any Wednesday or Saturday afternoon and evening would work for her to take Bertie. Then she headed over to Gerald and Jess's to give them their lessons. By the time she'd returned home to get ready to hold the Meetup, her sister had texted back, "Wednesday afternoon/evening good?"

She wrote back, "Yes!" and began to think of things she and Bertie could do together with Pooh-bah on Wednesday. In the fifteen minutes she had before the Meetup started, she dashed over to the bakery down the street and picked up two slices of pumpkin cheesecake to bring to Thomas's that night. Passing the wine store on her way, she dipped in, buying a bottle of red for their dinner.

The Meetup had some new members that evening who replaced a few who'd dropped out of the picture, although Nan knew all of this was par for the course with Meetups. Paige and Bruce were still hanging on and steadily sharing their progress with the group. Todd came for the latter part of the meeting, and a new member, an elderly woman named Daphne from Santa Clara, shared excerpts from a long piece for voice and guitar that she'd been composing for a few years now. Overall, Nan was pleased with their session.

Nan packed a few athletic clothes in her bag for that night at Thomas's, figuring that the next day she might run around Lake Merritt. She couldn't believe that after only a five hours' absence from him, she could be so excited to see him again.

"You sure this is all right?" she asked, standing hesitantly in front of the open passenger door to Thomas's BMW.

"Get in." He smiled. "I notice you're not nervous this time."

"Should I be?" she asked playfully.

"Always." He managed to arch his eyebrow in an at-once menacing and sprightly way.

"Seriously, how practical—or feasible—is it to keep me guessing all the time?" She laughed.

"To me, that question in itself constitutes challenge enough," he retorted. "I gladly take up your gauntlet." As if to seal their agreement, he placed his hand boldly on her upper thigh, a gesture that never failed to thrill her to her core.

She shook her head, smiling. "And you're not a bit tired from last night?"

"I feel more invigorated than I have for a long time. Are you?"

"You stole the words from out of my mouth."

"Good. I plan to put you through the paces tonight."

They had a succulent dinner of steak, potato slivers fried in olive oil (Thomas said this was one of his specialties), and baby kale salad with ricotta salata cheese, pomegranate seeds, and roasted pumpkin seeds. Nan's wine came in handy, and after they had finished dinner, she produced the pumpkin cheesecake.

"I know neither you nor I is a huge dessert person, but I strongly encourage you to make an exception for this—it's amazing," she avouched.

He poured some armagnac to go with the cheesecake.

"This really does feel like a vacation," she said as they sipped their armagnac and took bites of the cheesecake.

"Maybe it's something we both needed more than we knew," he offered.

"The pandemic has certainly got a lot of people out of whack where vacation is concerned," she remarked. "Many people I've talked to don't feel as if they've had a real vacation in a long time. Others feel as if the whole thing has been this weird limbo that's half-vacation and half-work, and neither feels right. And no one is taking their 'vacation' at the normal times."

He nodded. "A lot of customers at the bar discuss plans they have to travel somewhere eventually, and it sounds a little like inmates planning the things they're going to do once they get out of prison."

"Let's hope the time frame is a little better for the people at your bar."

Nan had put on a playlist of classic jazz from her iPod, and an old tune that she loved, a two-step from the 1930s, began to play. She stood and extended her arms in invitation to Thomas to dance with her. He stood with an obliging smile, and she showed him the basic steps. She didn't really care when he faltered, since she was in his arms, and she couldn't imagine anyone's arms she'd rather be in. They ended the number laughing, because either Thomas had stepped on her foot or she had stepped on his. She leaned into him and breathed in his Thomas-scent for the tenth time in a day's span. Nan realized then and there that she was a smell-and-sound woman.

"I wish I had a piano here so you could play for me."

"Next time we're at mine—after the ants are all gone—I'll play for you," she promised.

She noticed then that a wild glint had appeared in his eye, and she prepared herself for whatever was to come next. He stood directly in front of her, and she found herself slowly backing up towards the wall of the dining room—a strange déjà-vu from something she had imagined or dreamed in the last few weeks. When he had her standing against the wall, he placed his forearms to either side of her shoulders, effectively hemming her in. She muttered self-mockingly, "I'm afraid as prey goes, I'm not very smart."

A glimmer of a smile played about his lips as he met her eyes. "You're smart enough not to fight when you know you'll lose."

"Is that a challenge?" she asked, feeling at once emboldened and turned on.

Maddeningly, he said nothing, but held her gaze. In a swift movement she raised her hands toward his arms to resist them, but he, anticipating this, just as quickly redirected her energy and pinioned her arms behind her torso. The entire time, he remained close enough to breathe into her ear—an effect more provoking than all the rest of his actions. His calm, focused control contrasted strikingly with her scattered confusion, and she somehow felt that beneath it all he was goading her. She could sense the weight of him closing in on her against the wall, and, most frustrating of all, her loins began to exhibit deep pleasure at their assurance of his victory. Betrayed by her own body, and with nowhere to turn, she met his eyes defiantly. Perceiving her defiance, he pressed his lips passionately to hers, and she found her body and mind betraying her at once as she returned his kiss with corresponding passion. Their entangling tongues told the story of their power imbalance: his tongue would probe, draw hers out, dance and twist with it, and then withdraw just as she began to let herself go. Then his cocksure tongue would lingeringly trace the outside of her lips as if possessing them, before it re-entered her mouth to explore again. She lost all perception of time as his kiss slowly drove her from struggle to surrender.

When he pulled away slightly and the fog of desire cleared a bit, she began to recall her original purpose. She pushed against his hands behind her back, but found he was indeed stronger—or more determined. She writhed with her hips and attempted to move each of her legs outward. But he had closed them in completely with his own hips and thighs. She examined the strong deltoid muscles that confined her torso, and he, seeing her do this, said, "Yes, you could bite them. But it will be a great pleasure to me not only to bear those bites without flinching, but to have the marks as a lasting reminder of you."

She searched into the darkened recesses of his eyes and saw not just
a predatory gleam but a playful dare which instantly recalled her to the
delights of this situation and distracted her from its hazards.

"You're used to winning, aren't you?"

"Where you're concerned, it's always worth it," he replied, claiming
her neck with a kiss.

"How could you be so sure I would enjoy this?"

"I didn't," he confessed. "It was part of the ongoing experiment."

"You're quite a gambler," she said admiringly.

"You're learning as much about me as I am about you."

Chapter Twelve

Their foreplay—both verbal and physical—had Nan in a constant state of pressing desire for Thomas. Whenever he wanted her, she was his for the taking. That entire night they had a few small intervals of sleep punctuated by their bouts of lovemaking. Throughout, she preserved in the deep recesses of her brain the imprint of his dominance and boldness, his resolution and strong will. Her imagination was entirely complicit in cultivating the reality they were building between them, and that reality turned her on to an unsurpassed degree. She could see that the more she came to trust Thomas, the more freely and expansively they could both develop the possibilities of their "ongoing experiment," as he called it.

By morning they had fallen into a deep slumber at last, Nan's head resting in the space between Thomas's upper arm and pectorals, her arm draped across his chest, and her leg wrapped over his. She awoke eventually to find him gazing at her and a smile hovering over his lips.

"What is it?" She yawned.

"You fit me so well. In every sense."

"You don't know half of my faults," she objected.

"Try me." He moved his finger teasingly along the edge of her waist down to her hip. "If you can name one fault I don't know about yet, I'll let you choose our next dangerous scenario."

"Oh, otherwise you get to keep calling the shots?" She traced her fingers over the stubble of his jaw line. "I don't agree to that."

"You don't have to agree—nor would I want you to. But my offer still stands. Go ahead and see if you can surprise me."

"For one, I'm fiscally unsavvy. I make all the wrong sorts of decisions financially and dig myself into deeper holes every month."

He shrugged. "I surmised as much. So you're not the person to turn to when accounts need balancing. Lucky for you, I'm pretty good at that sort of thing."

"I've struggled in the past with being discreet on behalf of myself and others."

He nodded. "You're more honest than anyone I've ever met. I'm sure some indiscretion comes with the territory. But that doesn't surprise me either."

"I avoid almost all confrontation—except when I'm manic."

"I knew this about you already. You have a gentle spirit and a good nature. Those are the obverse of being non-confrontational."

"I go to the extreme in most good things—food, drink, exercise, sex, apparently."

"Yes, I won't deny your extremism. But yet again, that has a flip-side of passion, focus, and intensity—all good qualities."

"You know about my perfectionism and my inability to compromise. What you may not realize is that these traits mean I hold not just myself but anyone I'm close to the highest standards. And when I'm disappointed in my expectations, or when someone close shows vulnerability, I have a hard time accepting it."

"How does that work when you're teaching?"

She nodded grimly. "You've hit the nail on the head. During my last year at Drummond, my obsession over details and my impossible standards infected my teaching."

"Of course, even that flaw has its good side—I like teachers who hold me up to a high mark."

"Not everyone does, though," she said. "And they're right. Students should be able to come to a course with a variety of needs and have them be met. And grading according to a Platonic ideal helps no one."

"Well, you still haven't named something I hadn't deduced already. And in case you had hopes of scaring me away, you've accomplished the reverse."

"One more thing, however—not really a fault so much as a weakness," she added. "I'm fighting constantly to feel more secure in my own body."

"How can that be?" He seemed genuinely taken aback.

"My parents raised us all to think constantly about the body and how to keep from gaining weight while indulging in food and drink. Dieting and nutrition were the main topic of conversation the whole time I was growing up. In addition to this inherited familial obsession with body image, I have bigger bones than the average woman, and it's been a lifelong battle for me to come to accept that as my own particular form of beauty."

"That doesn't sound like a weakness so much as one more testimony to your strength." He kissed her tenderly on her forehead. "Now I find

I have a new project in addition to our joint seduction—and that is to help make you know how beautiful you are."

She ran her fingers through his thick hair. "Why don't you name some of your faults, and I'll say if any of them comes as unexpected."

He chuckled. "Where to start? There are so many. First and foremost, I'm not nearly as ambitious as I could be. I'm content with the comic side of life we discussed the other night at the party—although I know if I'd set my mind to something more serious early on, I could've accomplished it."

She placed her fingers on his lips. "Part of your comic brilliance is the way you're happy with your decision. Personally, I'm glad you chose the former and not the latter."

"I get easily impatient with idiots, and can't stand when people waste my time," he continued. "And I'm not always good at hiding that fact."

"You mean you have an occasional outburst of temper. But I sense that you're the sort to pick your battles carefully," she countered. "And, after all, having your time wasted is no small matter."

"Even more than you, I like—I need—to have control over things, from the smallest details to the bigger ones. Hence my slowness to open the restaurant portion of the bar. I couldn't stand to lose control of my place and upset the careful balance I'd achieved there."

"I could tell you were nervous about that when you first told me what would be involved. But I can't exactly fault your love of control—I thrive on that." She smiled mischievously.

He rested his hand comfortably on her hip and went on. "I'm more selfish than perhaps you're aware of. It was a calculated decision on my part to move out here when all my family are located on the Eastern seaboard. I didn't want my life to be defined by them, however much I love them."

She shook her head. "You're the youngest child—I'm sure had you stayed you would've felt more pressure than the rest of your siblings to mold your life to everyone else's. I'm afraid I can only respect your independence and strong sense of purpose. No, you haven't convinced me yet of any unredeemable flaws. Perhaps you and I are just looking at each other through rose-colored glasses."

His hand stirred to more intimate regions, and she groaned from the movement. "Now that you know the worst of me," he cajoled, "and haven't yet rejected me, I think that strengthens my position all the more."

"I may reject you yet. You can also be too arrogant."

But by then he had silenced her words with a kiss, as his hand knowingly roamed her most sensitive parts. She lost her train of thought entirely, lost all power of reason, and melded her body with his.

In the late morning, she convinced him to go on a run with her around the lake. Afterwards, she asked him to tag on some hills to the route so that they'd get more mileage and more resistance, and he led her first up to Piedmont Park, and then to the rose garden. When they'd finished those, she examined her odometer. "Can you come up with another two miles to make it a full ten?"

He burst out laughing. "I'll show you a climb that should do you in." He led her to Trestle Glen, behind the Lakeshore Trader Joe's. "You go ahead and climb as far as you want. I'll get a coffee at Peet's and wait here."

Nan thought, "How bad can this be?" But after about half a mile, she began to see what Thomas had meant. The climb rapidly became steep, and it seemed to have no end in sight. After venturing along this hill for exactly one mile, she turned around with relief, sweaty and out of breath, and flew down to where she'd left Thomas.

He was sitting on a bench sipping coffee in the sun taking in the bustling activity on Lakeshore. She crept up behind him and was just about to put her hands over his eyes, when he turned his head slightly. "How was it?"

In her bafflement, she had to laugh. "How did you hear me in all this noise?"

"I saw you out of the corner of my eye a minute ago, and figured you'd be here right about now."

"I was going to surprise you."

"I guess it's just a habit from riding the bike—being aware of your full surroundings at all times."

She took a seat beside him. "You sound like someone trained in the military."

"In that one respect, I might actually have done pretty well in the military. Otherwise, I'm sure I wouldn't have even survived basic training."

"Too individualistic?"

"That, and too independent-minded."

"I once heard a saying that the world is made up of two kinds of people—those who are willing to lead, and those who refuse to be led. I gather you're in the latter group?"

He nodded. "I'll admit it's not a very helpful position to take in society."

"That makes you like my Quaker family. But as a group who all refuse to be led, they still manage to contribute a lot to the community."

"A community doesn't have the kinds of hierarchies that society does," he said. "From April through October, I volunteer at an Oakland community center and show kids how to fix bikes and sharpen their bike-handling skills. That kind of leadership appeals to me, and the kids—who all want to be there—keep their own autonomy through the entire process."

"How do you obtain bikes and tools?" she asked.

"A local bike shop sponsors it, and we raise additional contributions. By the end of each six-week program, each kid has a bike, a knowledge of essential repairs, and a good command of riding basics."

"What age group?"

"Seven to seventeen years old."

"You must really like kids."

"I especially enjoy the younger ones, I admit."

She tilted her head up and closed her eyes to soak in the sun. "I guess I'm learning to appreciate pre-teens more as I take on younger students. But I usually skew towards the 14- to 18-year-old range myself."

"And how do you feel about having children of your own?" he asked.

She was startled at his forthrightness. His question stirred up memories of a year and a half ago, just before her thirty-fourth birthday, when her loneliness had taken a surprising turn. She had suddenly felt the overwhelming need to consult with the fertility department at her medical care facility to find out the chances of her still being able to reproduce, and the options available for doing so. She had had a blood test done to check the viability of her eggs and, after receiving an enthusiastic report from her doctor, had proceeded to have an ultrasound to assess the quantity of her eggs. This too had yielded favorable results. In the month that followed, Bess—who was five years older and had given birth to Bertie at Nan's current age— had told Nan all about the processes she had looked into, as well as the methods friends and acquaintances had used, to get pregnant.

After Nan had invested this initial thought, time, and money into the matter of having a child, almost as quickly as the urge had come upon her, it subsided. She reasoned that there could still be a partner out there for her, that it was far better to raise a child with a partner, and that financial straits militated against attempting to do so now. She also endorsed the possibility of eventually adopting a child, which seemed an exciting adventure in its own right. As practicality overrode her loneliness, she relaxed into the idea of using the time during the pandemic to create in other ways, namely, through art and music, and vowed eventually to start dating again.

Now she replied, "I've always wanted to have children—whether biologically my own or adopted. What about you?"

"I'm open to the possibility, with the right person." He took her hand, bringing it up to his lips and kissing it.

In the ensuing silence, Nan felt incredibly shy, and she shifted the topic of conversation back to Thomas's volunteer work.

"Your bike workshop sounds amazing. I'm still trying to figure out how to do something like that in music education. I've been thinking of giving some free music appreciation and history classes at the retirement home, and opening them up to the community."

"That sounds like something people would be interested in."

"I'll ask Fannie this Friday if we might start that up in the New Year. Thanks for the additional motivation."

They jogged back to his house, showered, changed, and had a sandwich for lunch. He dropped her off at her place on his way to the bar.

She dipped her head into the car after stepping out. "If I haven't already overstayed my welcome, I could come to your place or your bar on my bike later after the chamber group meets."

"Come to the bar, and I can pack your bike into the back seat of my car after I've finished. That way, tomorrow we can head out directly into the hills from my place."

"Hopefully my landlord will bring in the exterminator tomorrow, and I can leave you in peace."

"You know me well enough by now to know I don't relish peace. I hope I never do."

She laughed. "Good luck with the interviews this afternoon."

Joanna was on point in every respect that afternoon: she'd worked on the fingering they had together decided worked best for her own hands; she really leaned into the dynamics; and her phrasing was expressive and musical.

"I love how you've developed this piece, Joanna," said Nan. "Fauré would be pleased to hear you interpret it."

Joanna beamed modestly. "My grandmother's really excited about the recital. She always wanted to learn piano but never got the chance. And she hasn't yet heard me practice it—unlike my parents, who hear it all the time."

"Well, even to your parents this piece will sound new, because we'll be playing it together—so they'll hear you with the Secundo part at last."

After she'd given her lessons to Joanna and Allen, Nan had some time to ship off a few more CDs that had sold on eBay. She was dismayed to see that her checking account might be overdrawn this month, if she wasn't able to scrape up $260 more in cash. She decided to sell something big on Craigslist before the middle of the month to cover the minimum payments and other bills that would be coming due around the fifteenth. She warmed up on the piano with some Schubert impromptus and Brahms pieces, and then rode over to Jules's house for the trio's last practice session before their concert on Saturday.

Jules said he might place some chairs inside for those guests who felt too cold to be out on the patio. "It's going to be on the colder side this Saturday. I have one heating lamp, but that's not going to go far."

"I'll bring more chairs," Céline offered.

"What's most helpful for me to bring?" asked Nan.

"I suppose we can always do with more wine." Jules smiled. "All in all, this should be on the same level of intimacy that Beethoven conceived of for his chamber pieces. I can say a few words beforehand about the trio. If either of you has anything particular you want me to add, let me know."

After they had played through the piece once fully and then gone back over a few tricky parts, Céline said, "Nan, I forgot to say, if you want a page turner, Cécile has offered to do it."

"Oh, thank you! I'll take her up on that."

They had a glass of wine together as they discussed the Schumann quintet for the New Year.

"Despina's friend Gregorio said we could have his space Saturday, March 19, if that's not too soon," Jules announced. "This will give us an even better recording opportunity, and more seating space."

"And Despina and Paul love our name, 'The Perseid Quintet,'" Céline added. "So we can put that on the program."

"Shall we prepare a few other pieces, to make it worth the audience's while?" Nan proposed. "Maybe some solo pieces for violin, cello, or viola and piano?"

"I'd love that." Céline sounded vibrant. "But we don't want to overtax you."

"I never can get enough of playing with others." Nan sipped her wine. "I was going to suggest that we start a YouTube channel and put up videos of our recorded sessions. That is, if it's all right with all of you."

Jules and Céline nodded enthusiastically.

"What a great idea," Jules remarked. "It's something I've always meant to do, but never got around to."

They agreed to broach this with Despina and Paul, before brainstorming on pieces they could include in the March concert.

"It's wonderful to have things to look forward to after the New Year," Céline said in a sprightly tone.

"Now we just have to get *to* the New Year." Jules's smile reinforced his irony.

♪ ♪ ♪ ♪

Later that night, when Nan brought her bike into Thomas's bar, Benny waved her in summarily. Dan Fisher was standing at the bar talking to some men in their fifties and sixties, and Thomas was nowhere to be seen. Elian nodded at Nan and gestured for her to take a seat at the bar. Nan took a place a few seats down from where Dan was regaling the other men with what seemed to be holiday-themed jokes.

Dan smiled briefly at her, and then said, "So when we got to the party, everyone was making merry. Then Mary left and we all jumped for Joy. Joy got mad and left, and a lady jumped out of the party cake. We all had a piece." He paused. "The cake wasn't bad either."

The group roared. Then one of the other men said, "Yeah . . . why doesn't Santa have any kids?" He paused for a beat. "Because he only comes once a year."

Dan upped the ante and told an even raunchier joke: "Three garbage men are driving around a neighborhood collecting their Christmas Eve tips. They come to a nice quiet cul-de-sac where they pull up in front of a cozy house. One of the garbage men goes up to the door and rings the bell. A sultry housewife in revealing clothes opens the door. 'Come in,' she says invitingly. 'I've been expecting you.' She takes him upstairs and has her wicked way with him. The garbage man comes out looking pretty pleased, and tells his two companions what's just happened. The second garbage man decides to go and try his luck. The housewife opens the door to him in her revealing lingerie and invites him upstairs, as if she's been expecting him, and she has her wicked way with him. When the garbage man comes out looking just as happy as his companion, the driver decides to go and try his luck. He rings the bell, and the housewife comes and hands him a five-dollar bill. The driver asks her disappointedly, 'But why did you seduce my two companions and give me five dollars?' 'Oh that,' says the wife. 'That was my husband. When you all parked in front he told me to "just give the driver five bucks and fuck the rest."'"

Nan indulged in a quiet laugh behind her hand, and then asked Elian for some water, which he immediately gave her. He seemed apologetic. "Can I get you a beer or something? Thomas should be back any moment. He wanted to get to Target before they closed."

"I'd love a glass of sauvignon blanc, if you have it," said Nan.

As she sipped her wine and pulled out her Nancy Mitford novel, the group of men seemed to sense her presence and they shifted gears to other, less risqué topics, including a motorcycle that one of the men's friends was trying to sell.

"He's never gonna get rid of it if he's crashed on it," said Dan wisely.

"The crash wasn't that bad. He took a hairpin turn a little too fast and there was gravel on the side of the road. He and the machine just got a little scuffed up, that's all."

"I know someone who's looking for a bike," said one of the men. "I'll tell him about it. How much is he asking?"

Soon they all asked for more beers, and Elian became even busier drawing their drinks.

Dan came over to Nan. "I know you. You're the gal who came here a while ago with the knock-out legs. I think you're Thomas's new girlfriend, is that right?"

Nan flushed to hear herself called that. "I guess so."

"Don't you know?" Dan waggled his eyebrows suggestively, and Nan couldn't help laughing. "Thomas is a good guy. I've known him ever since he worked for Harry, long time ago. He gets along with all types. But he's got some crazy notions. He's gonna change everything here, put us out for a while, start serving food." Dan shook his head. "Helluva time to change this place. I hope he knows what he's doing."

Nan said, "He's thought it through pretty carefully. Wait till you see the website."

"Everyone's got a website these days." Dan was thoroughly unimpressed. "My dentist has a website. Nothing to put up there except the price of teeth. You spend so much time looking at websites you don't go to the place itself anymore. Now, I can't tell you how many times I've seen young couples come in here, order a drink, sit across from each other, and play on their phones all night. They might as well be in different time zones as at the same table. Never exchange a word. And everybody has to have a 'Wi-Fi password' now when they come to a restaurant. Why? What's so important that you have to have high-speed internet while you eat your pancakes?" Dan took a sip of his beer as he warmed to his subject. "I saw a lady the other day at the bank, young like yourself, plunked her kid down on the floor with one of those tablet thingies and told him to play on it while she got money. Kids these days are being raised by Facebook and YouTube."

Nan couldn't help agreeing. "There's something to that."

"Now, I'm not saying that our generation didn't have its faults too. People used to dump trash everywhere. Now you've got your different holes for cans and paper and whatnot. They hold your ten cents at the store until you bring your bottle back. A lot more work for everyone, if you ask me. But maybe that's the whole point."

"What sort of work do you do?" asked Nan.

"I'm retired," said Dan. "I used to own an auto repair shop down that way. We specialized in fixing German-mades and older models. I sold the place last year to the son of a friend. Seemed like as good a time as any to call it quits."

"What do you plan to do now?"

"Enjoy life!" Dan finished his beer. "I've got grandkids in the area. I have friends who also like fishing. And my son's family is taking me on a cruise next year around Greece."

"That sounds like a pretty nice life."

"I can't complain. My wife died a few years ago. I wish she could've enjoyed all this with me." Dan looked sad. "She always wanted to travel more. And this was her favorite time of year, the holidays." He sighed, and then pulled himself up short as he winked at Nan. "But there's your man coming in the door now. Thomas, you've got a lovely girl here. I've been talking her head off."

"I'm sure you have." Thomas grinned. "Haven't you got anywhere else to be, Dan?"

"I'm your best customer—you know you're lucky to have me."

"You're good for bringing in other business." Thomas spoke in measured tones. "I tolerate you for that reason alone."

"You see what sort of gratitude I get?" Dan turned to Nan. "Forty some years I've been coming here."

"At the rate you're going, you'll come for another forty years." Thomas stepped behind the bar to pick up some empty trays.

"I'll take that as a compliment on my excellent health. Elian, lemme have another one."

"And by the way, you drink too much," Thomas added.

Dan shot him a pretend-dirty look and began to engage Elian in conversation.

Thomas gave Nan a kiss. "Let me guess, there were dirty jokes."

"A few." She took a sip of wine. "But they were actually pretty funny."

"I think I've heard them all in my time," said Thomas. "It's an occupational hazard of being a bar owner that you stop finding jokes funny because you know all the punch lines."

"Are the rest of those guys regulars too?" She gestured over to the group of men who'd been talking to Dan earlier.

"Yeah, most of them. They're the weekday crowd. A lot of them don't like coming on weekends because of the 'riff-raff,' as they call the younger, newer crowds."

"How did the interviews go?"

"We hired a few people for the kitchen and wait staff. Only a few more positions to fill and I can breathe a little more easily."

Nan helped him move chairs back to the emptier tables and clear up glasses Elian hadn't been able to get to. Thomas took out some recycling and put a load of rinsed glasses into the dishwasher. Eventually, he nodded over to Elian. "You okay for the rest?"

Elian smiled. "No problem. See you tomorrow."

Thomas took off Nan's front wheel and carefully placed her bike in his back seat, nestling the dropouts in an old blanket. "I don't care about the seat," he explained as she watched him. "I just don't want to scuff your frame."

When they entered his living room, she sighed contentedly. It felt like home by now, so much had happened here over the last forty-eight hours.

"If it's okay, I would love to take a bath tonight," she said.

"Please do." He spread his arms wide. "And then feel free to walk around naked afterwards. I can sit or lie wherever you want me to for best effect."

She laughed. "This is day three, and you're really not tired?"

"I'm just warming up." He stepped to her with a smile, enlacing her waist. "I can give you a night off, if you want."

"Well, that's very considerate of you," she said stoutly. "But I'm no weakling. I can take whatever you dish out."

"That's what I thought." He kissed the tip of her nose tenderly. "I want to talk to you about something. But first, take your bath."

She tilted her head with puzzlement. "Is it something bad?"

"Far from it. The opposite."

"Will you come in while I take my bath and tell me then?"

"Gladly."

She drew her bath and sank luxuriantly into its steamy heat. She hadn't taken a bath since the last time she'd been to Minneapolis to stay with her dad, so this was a real treat. She closed her eyes and breathed deeply in and out. *This* was true heaven. After a while, Thomas knocked on the door, which was ajar, and entered.

"Are you decent?" he teased.

"You didn't have any bubble bath, so I'm afraid you can see everything," she said with little shame. "But then, you already have."

He smiled wickedly. "I have indeed." He admired her frankly for a few moments as she rubbed soap over her body. "You know, I have an idea about something we could do; I've thought about it ever since I met you, to be honest."

"Oh?" She felt a tingling in her lower regions, but she continued to rinse her face and shoulders.

"What would you think about some more uninhibited scenarios between us?"

"I thought we *had* been pretty uninhibited." She laughed.

"Well, I have some ideas about ways we could explore our dynamic more explicitly."

"Can you tell me about them in detail? I feel as if I prefer that to the subtler game we played last night—even though that ended up being sexy as hell."

He propped his elbows on his knees. "You're teaching me to communicate more openly about things like this. My idea is to use silk scarves and tie you down. I know it sounds clichéd, but I don't think it will end up being so when we do it. Nothing you and I do is clichéd."

The blood rushed to her groin as he described what he had in mind more fully. She nodded. "I think I'd like that. I'll leave it to you."

"So you trust me?"

"I do."

"We don't have to do it tonight. I need to get a few more things. But some time soon."

"Okay, well, just the idea of what you said makes me horny enough for the entire night. So I think it would be redundant to actually *do* all of that tonight."

He smiled. "When you get out of that tub, don't put anything on. I never thought of the bath as foreplay, but it's working with you."

When she stepped out, he helped dry her with the towel, standing just behind her. She felt his erection, and found herself turned on by his excitement as well as the idea of what they were eventually going to try together.

She had thought she was incredibly tired. But once they had started to move slowly out of the bathroom, Thomas guiding her from behind her with his hands around her pelvis, she felt fully renewed. He led her to the counter in the kitchen, where he picked her up by the hips and seated her on the edge. She laughed in disbelief. "I don't think of myself as light by any means. But somehow you manage to make me feel light."

"Your musculature is perfect. It's beyond most people's wildest dreams." He kissed her breast. "And then there are the other parts—the non-muscular ones . . . " He trailed off in rapture, and she sighed deeply as he held her steady and touched her in the right places.

She loved the offbeat, indecent feel of doing it in the kitchen. By the time they had fallen into his bed, she thought she could sleep forever. She shut her eyes, and neither of them awoke until 8 the next morning.

As they made breakfast, he asked, "Where are we headed to this morning?"

She thought for a second. "You want to do the Paradise Loop in Marin?"

"All 65 miles?" He sounded skeptical.

"I need to redeem myself in Marin County," she pointed out. "Remember my last foray there."

"All right. Do we get lunch in Tiburon?"

"Yes!" Nan hated packing lunches for bike rides, but loved engaging with locals at cafés to procure her food.

He looked at his watch. "Okay. Let's get going. You have to be back by 4 for Bertie."

She hadn't forgotten. They made good time. They had sandwiches and coffee in Tiburon midway through the ride and, as usual, the climb up Camino Alto in Mill Valley seemed tame in comparison to the hills of the East Bay. Thomas got a flat at the base of the Richmond Bridge, causing Nan to comment, "This place is cursed."

She impressed him with the speed with which she changed his flat.

"From now on, I'm just going to sit back and let you fix all the flats," he said appreciatively.

"I don't think men are as good at feeling for the glass in the tire as women are," she joked. "It takes a sensitive touch."

They took turns drafting one another to combat the headwind that met them all along the bike trail on the east shore of the Bay. They made it back to Thomas's by 3:15.

"You get a thrill from cutting things close, I see." He arched a playful eyebrow.

"It always feels as if I've made the most of time that way," she confessed.

They showered and dressed in record time, he put her bike in the back seat, and drove her to her place on his way to the bar.

"You're sure you don't want me to take you to Bertie's school?" he asked.

"No, it's less than a mile up this way. I'll go on my bike and then walk with him and Pooh-bah back here."

"So the exterminator's coming this afternoon?"

"The landlord said they'd come and spray the wall at 4—I'll keep Bertie and Pooh-bah out playing for a while so none of us breathes in the fumes. Let's hope this works."

"Well, you know I'll be happy to have you at my place again, anytime."

"Thank you. For everything."

They kissed lingeringly, and he cradled the back of her neck delicately in his fingertips. Then in a flash he had stepped back into the car, put it in gear, and pulled away. Nan hurried into her apartment, deposited her backpack, and took off to meet Bertie.

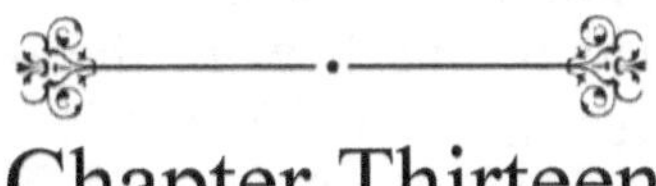

Chapter Thirteen

Bertie seemed to have grown an inch since Nan had last played with him. They took Pooh-bah for a walk up the hillside to the track where Nan often ran. UC Berkeley, which owned the track, had dumped huge piles of dirt on the edge of it to distribute at places that were eroding. Pooh-bah gleefully hollowed out tunnels in the dirt using his broad dachshund's paws, and Bertie shouted in triumph when he'd completed three such pits. They then proceeded down the other side of the hill towards a grassy meadow where they could throw the ball for Pooh-bah. They took Domingo Avenue, passing the Claremont Hotel tennis courts on their left and Peet's coffee to their right.

"There's Morvan again," said Bertie. "Are you going to say hi?"

Nan was caught utterly off-guard. The last person she'd been thinking of was Oliver. But there he was, smiling over at them from a sunny bench to the side of Peet's. Nan smiled at him, struck more than ever by his strong resemblance to Bertie. To her dismay, he got up and came over towards them. Nan cast about looking for a way she could distance Bertie from the scene, but Pooh-bah, held on the leash by Bertie, strained to greet Oliver as well.

"How are you, young man?" said Oliver to Bertie. "I'm Oliver."

"I'm Bertie, and this is Pooh-bah."

Oliver turned to Nan. "I haven't seen your sister at church for a while."

"She's been pretty busy." Nan tried to think of a polite exit strategy. She recognized the precariousness of her position, especially with Bertie standing right there. Oliver was studying Bertie more acutely than she would have wished.

"Do you know my mom?" asked Bertie.

"She's a good friend—or was a good friend." Oliver spoke in a wistful tone. Then he said to Nan, "If you could tell her that I know why she's been staying away from church, and I know about Bertie, but I won't make any claim . . . please let her know that the whole thing is a matter of conscience for me."

Nan listened, memorizing every phrase so that she wouldn't get any of it wrong when she passed the message on to Bess later. Oliver saluted Bertie and turned on his heel in the opposite direction from where Nan and Bertie were headed.

Unfortunately, Bertie was full of questions as they walked. "What did he mean when he said 'I know about Bertie'?"

"I think he meant he had seen you before—remember at Monkey Island?" Nan faltered.

"Why has my mom been staying away from church?"

"She wants to spend more time with you on Sundays," she hedged.

"What does he mean when he says he 'won't make any claim'?"

"I'm not sure." This time she didn't have to equivocate, since she didn't fully understand what Oliver meant.

"You said his name was Morvan."

"That's his last name."

"Oh. What's a 'matter of conscience'?"

She laughed. "You have a good ear and memory, Bertie. You'll make a fine musician someday soon. I think 'matter of conscience' means you're committed to doing things right, acting for the good of the people involved."

Bertie nodded. "I like Morvan—Oliver. So does Pooh-bah."

"That's always a good sign, isn't it? When Pooh-bah likes someone."

They tossed the ball for Pooh-bah in the meadow for 45 minutes, and when both Bertie and Pooh-bah were sufficiently tired out, Nan led them back to her place. She looked to the side of her door inside, and sure enough, milky white spray covered the wall and hundreds of dead ants bore witness to its efficacy. No more ants marched along the walls or counter. She went downstairs and saw that activity had been considerably slowed there as well. Only a few solitary scouts remained, and the loss of their colony seemed to have quelled their spirits. She heaved a sigh of relief and came back upstairs.

"What would you two like for dinner?"

"Pooh-bah likes chicken. So do I."

Nan decided to make up some plain chicken and broccoli for both of them, and to cook a spicy stir-fry for herself using the same ingredients plus a few other vegetables. They listened to the classical radio station while she cooked and asked Bertie about school, the books he'd been reading, and the games he was playing these days. He told her about a few friends at school who shared his love of Pokémon trading cards, and even spelled out these friends' names. Nan was impressed by his instinct to be so precise.

Pooh-bah lay curled up like a warm croissant in Nan's quilted jacket, his crackly didgeridoo sound of contentment activated each time Bertie or Nan caressed his belly. By the time Bess came by in her car at 9, they were immersed in a game of checkers using a book of board games Nan kept on her bookshelf.

As Bertie went downstairs to use the bathroom, Bess asked, "How was it?"

"He's such a darling, as usual." Nan then told Bess about Oliver and quoted his words verbatim.

"That doesn't comfort me a bit, to know that it's 'a matter of conscience' to him," Bess said with pique. "And what sort of claim does he think he could make? He doesn't know I can't find that paper. I think I'm going to attend church again this Sunday and set the record straight with him."

Nan kept strategically silent this time. She may no longer sense Oliver as a threat, but clearly Bess did. "How's Dennis doing?"

Her sister's frown turned to a smile. "He's great. We went to dinner at a superb Thai place in Mountain View, not far from work. Then we had a long walk all around the downtown, which I've never really seen up close, all the years I've worked for the Institute. We're going to take a brief road trip to Nevada City, just before I come to Minneapolis for the holiday. Dennis has never been up to the foothills of the Sierras."

"That's wonderful. Will you stay in an Airbnb?"

"We'll stay in the vacation home of Dad's old friend, Doug Huston. I told Dad we were going up there, and he arranged to have Doug's daughter leave the keys for us."

"Wow, that'll be incredibly romantic. You guys should have snow up there this year. It's been really cold in the Sierras."

"Yeah, that's what I'm hoping for. Lots of nice fires in the fireplace, and when Bertie's gone to bed, we can canoodle a little." Bess smiled. "By the way, speaking of canoodling, I get the sense some of that's going on with you and Thomas—am I right?"

Nan flushed. "Yeah, it's gotten pretty serious with him."

"Oh! You've gone all the way," said Bess astutely. "I can tell from your face. Is this going to derail you from your composition and performances?"

"No! Not at all. I mean, so far I feel as if it's taken off a lot of the stress I had before we got involved. That's been a good thing, for writing, practicing, performing . . . everything." Nan blushed anew.

"Well, you'll have a hiatus at Christmas when we're in Minneapolis."

"Don't remind me," said Nan with a trace of dread. "I'm not looking forward to that."

"Don't let Dad know you're not. He's really excited about our coming."

"In theory, I love every second of what we do over the holidays with the family." Nan filled the kettle with water. "But this year, I'm going to be missing him all the time."

"Where will he be?"

"I think in Baltimore, with his sister's family."

"Well, you know what they say about absence . . ."

"I know." Nan smiled reluctantly. "But no one who was falling in love ever endorsed that saying."

"True," said Bess. "It does sound like something some old fogey invented to keep the youngsters in check."

"At least we can lose ourselves in Teddy's wedding preparations this year. That should help distract us from missing our men."

"And there's always Facetime," Bess pointed out.

After Nan had hugged them both goodbye, she saw that a text had just come in from Margot. "You up for a chat, love?"

She called Margot. "How is everything, dear? We haven't spoken since Saturday evening." Nan felt more than a little guilty, since she'd been so wrapped up in Thomas that she hadn't even texted Margot.

Margot said the date with Aaron had gone well on Saturday, and that they had mapped out ways of spending time together in the course of the weekdays. "But," she added, "he's canceled twice on our plans. The first was for us to go running together on Monday. The second was for us to have lunch at his job today. And now we have no further plan at all—not even for this weekend. I feel as if I'm doing all the heavy lifting, and he's getting more and more absorbed in work pure and simple. He got the interview with Google."

"Oh, wow. When is it?"

"Friday."

"Well, then I'm sure he'll be able to focus more on you after Friday."

"But why no plans for the weekend?"

"Give him a few more days. Maybe he just can't think until after the interview is over."

"I'm worried about what happens if the interview doesn't go well, and he doesn't get the job."

"From what you've told me about Aaron, it sounds as if he has a large enough fund of wit and good humor to weather that eventuality."

"I hope so," said Margot, half doubtful, half sanguine. "How's everything with Thomas?"

Nan told Margot a slightly abbreviated version of everything that had happened since Sunday. Margot let out a low whistle. "Wow. You guys are sizzling. When will you see him again?"

"I think on Saturday, when we go for a ride with two of his former cycling buddies."

"I'm looking forward to your concert on Saturday evening," said Margot.

"Do you want Thomas to pick you up at the 19th Street BART so he can take you to Jules's place?"

"That would be wonderful. Tell him thank you."

"I only offer up him and his car because he'll be right around the corner at his bar."

Margot laughed. "It's okay, Nan. I hear the love in your voice. No explanations needed."

Nan was surprised. "Bess read the same thing in my face. I guess I'm an open book where Thomas is concerned."

Thursday was full of activities Nan loved, and she sang the jazz song "That Old Devil Moon" to herself as she made breakfast in preparation for the day. She warmed up for two hours on the piano after breakfast, as she and Albert were going to be rehearsing the second movement of the Franck violin sonata, which featured a challenging part for the pianist. She was relieved that she'd moved her therapy appointment to the following Monday. She and Albert met for a full two hours in the practice room on campus and, as usual, Albert had the gift of gab; but Nan managed to keep him more or less focused on the piece. She made a mental note to work more intensely on Franck herself in upcoming weeks, since she didn't want to take the rippling second movement any more slowly than the intended "Allegro molto."

When they'd finished, she said goodbye to Albert and, as usual, jetted over to take Hershey for his walk. This time, she took him to Upper Rockridge, into the hilly neighborhood of the student whose mother was hosting the second recital in two weeks' time. Hershey was a natural-born hiker and looked, at the end of the walk, as if he could've continued another ten miles. Nan hugged him and played with him a little before returning home for her lessons with Melanie and Linda.

Nan had to reassure Linda that the stakes were very low for her in the upcoming recital, and that the only standard by which she was being measured was the preparation she put into the pieces.

"Remember," said Nan, recalling similar words Lilian had spoken to her years ago, "no matter how your performance goes in the actual recital, the audience, and you yourself, will hear the effort you've put in leading up to it. That's all that matters."

Linda nodded. "I'm 66 years old. I think at 26, or even 36, this wouldn't have fazed me a bit. But it gets harder to do things like this as you get older."

"Well, brava to you for tackling it anyway," said Nan, with heartfelt admiration.

After Linda's lesson, exhausted as she was, Nan rode over to the gym to collect the equipment for her boot camp. There were two new people waiting at the schoolyard, both eager to shed some Thanksgiving pounds with the rest of the group. The holiday season was always a good one for gyms, the rule holding for boot camps as well. Nan loved the challenge of working the playground and its equipment into the routines she devised. Tonight, inspired by Thomas's bar, she used a playlist of mostly Cuban music, cueing up the songs to encourage participants to alternate among fast, moderate, and slow paces in the workout. After they'd cooled down and the attendees had thanked Nan, some saying they'd loved it, others groaning and shaking their heads, Nan began to put everything back into the gym's bag.

"Can I help you with that?" said a soft baritone voice that registered first as a thrill in Nan's solar plexus and then zinged to her ears and brain in that order.

She looked up eagerly to find Thomas standing there, smiling.

"Where did you come from?" She beamed, utterly charmed by his tracking her down in order to surprise her.

"My car. Well, before that, my bar. I thought you might like to go out to dinner here on Piedmont Avenue."

"Won't I be under-dressed?" Nan looked down doubtfully at her workout clothes.

"I have a clean tee-shirt in the car for you to wear so you don't get cold. I also have an extra jacket. You know the Bay Area—no one here really cares about dressing up. Besides," he added in a lower voice, "you look gorgeous."

The last of the boot camp attendees called out his goodbye, and Thomas helped her put the rest of the gear away.

"They all look pretty worn out," he remarked. "You've clearly done your job and then some."

She met his amused glance. "They said they wanted to work off the stuffing and gravy."

"How does The Wolf sound to you?" She looked confused, and he clarified. "The restaurant down the block from here. They have excellent wines and cocktails, and their food menu is really original—part French, part American. They have stew."

"Oh, stew sounds heavenly. I just have to return this bag to the gym."

Now that they were alone in the schoolyard, Thomas drew her to him and kissed her. "I've missed you. It seems like longer than yesterday afternoon."

"It does," she agreed, closing her eyes and giving herself over to this beguiling moment. "A lot has happened since then."

They walked her e-bike down to the gym, dropped off the duffel bag of gym gear, and then headed back in the opposite direction, stopping in front of the restaurant.

She secured the bike, and Thomas wrapped his arm around her, feeling her shiver.

"Let's get those clothes for you. My car is right here."

They succeeded in obtaining a table in the crowded main room of the restaurant—"You're lucky," said the host, "we just had a cancellation"—and Nan went to the restroom to change into Thomas's shirt. When she came back, he was poring over the menus as if they were books of philosophy.

"What looks good?" She took a seat across from him.

"I don't know how you feel about pâté, but they have a duck liver mousse here."

"Mmm. I love liver anything. What else?"

"How does the octopus sound for a second starter?"

She nodded enthusiastically.

He continued. "And I think you said the seafood stew sounded good to you right now. What would you like for a second entrée?"

She considered the food menu. "I know we're getting a duck liver mousse, but can we try the honey-glazed duck breast? I can never have enough duck."

"That works for me," he said easily. "Are you in the mood for cocktails or wine?"

She examined the cocktail menu. "I see two drinks that look delicious. But if you've found any nice wines by the glass, I could go that direction too."

"How about a bottle?" he suggested. "I think we couldn't go wrong with this Piedmont white. It should pair nicely with all the dishes we've chosen."

As usual, being under Thomas's care in a restaurant felt like riding a ferry boat headed to an unfamiliar destination: the all-too-brief journey entailed nothing but pleasant views, comfortable passage, and lively conversations, the endpoint promising newness and fascination.

When Thomas had put in the complete order with the waiter, Nan asked, "What are we celebrating? It feels like a huge treat to come to a place like this." She glanced around at the beautiful wood paneling, the subtle decorative accents, the contented patrons who'd comfortably filled the place as if it were a familiar corner bistro.

"It would seem I've secured my entire staff for the new bar-restaurant. Not counting Luca, we'll have ten new employees come the New Year."

"That was fast!" Nan was duly impressed. "How did you manage it?"

"I couldn't have, not without Elian's and Luiz's help—not only did they put the word out to people they know and trust, but they vouched for me as an employer. So hopefully it'll be a win for everyone."

"Congratulations," Nan exclaimed as the waiter stepped over and opened their bottle of wine. After allowing Thomas to taste a bit, he poured two glasses for them, nodded, and left. They clinked glasses and Nan said, by way of a toast, "To The Chaser Bar and Restaurant. Long may it flourish."

Thomas half-smiled. "Let's just hope it's not a classic case of getting too big for one's britches. This Omicron variant may do me in yet."

"There's an old saying from Italy—you have to aim high in order to hit the target."

"I know from my experience with you how true that is."

Nan blushed. "I could take that in a few senses."

Thomas laughed. "You could, but I promise I meant it innocently enough."

∫ ∫ ∫ ∫

Over the appetizers, she asked after his sister Ellie.

He shook his head. "They convinced her to see a psychiatrist and go to group therapy. But it's going to be a long, rough road ahead. If she takes her medications regularly, she should start seeing some changes in a few weeks' time. But she needs to keep going to therapy and practice all the exercises they suggest. And then there's the clearing out of the house. Ken has ordered a dumpster to be placed in front of the house for the next week, and everyone's pitching in for a few hours here and there. It's complicated, because there are the usual keep, donate, and throw away piles, but the donate piles are the largest. A lot of the stuff is new in unopened packages, and it needs to be fully sorted through to be taken to various charity stores." Thomas ran his hands through his hair. "Honestly, I feel pretty guilty being out here and not helping them."

"Will you go to Annie's for Christmas?"

"Yeah. I'll take an early-morning flight on Christmas day and come back late on the twenty-eighth, so it's not a long trip by any means. But at least I'll get to see Annie, Leo, and Martin—my nephew."

Thomas asked after Bess and Bertie. After careful consideration, Nan had decided to share with Thomas the situation going on with Oliver. She told him the whole story from its beginning a few weeks ago to the present.

"So now my sister is going to talk to him again after church on Sunday. I feel sorry for him. I think she should just come clean and trust that he's the same morally upright man she chose to be Bertie's father seven years ago."

"You refer to him as 'Bertie's father,'" said Thomas. "But it sounds as if your sister wants him to remain solely his sperm donor—at least for the time being. Whatever Oliver wants is immaterial. Your sister is probably protecting herself and Bertie as best she can."

"I guess so. I just wish he wouldn't keep using me as an intermediary. I'm lousy at it, because I see his point of view better than Bess's."

"Your sister sounds like a force to be reckoned with—like you."

"You'll meet her on Saturday. Oh, is it all right if you take Margot from the BART at 19th Street? I'm afraid I volunteered you already . . ."

He nodded. "Glad to do it. Have her text me when her train gets in."

Their entrées arrived and, for a few minutes they gave themselves over to the rich smells and tastes of the seafood stew and the duck breast. Both of them were hungry from long, active days, and they set about the dishes with gusto.

Nan cut up a piece of cod. "I gather there's quite a clash going on in Austria now between the draconian government, on the one hand, and the libertarians, on the other—and in between, there's not enough vaccinating going on."

"My friend Stefan thinks the government's decision to mandate vaccinations by February and to lock down the country again is the only way they can curb the huge death rates. Apparently there's just too much disinformation being spread about vaccines, and it's having a lethal effect."

"Does he still live in Austria?"

"Yes, in Vienna. He's an assignment editor for the *Wiener Zeitung*."

"Wow, that's like *The Times* in England, right?"

"Yeah. If you can believe it, though, Stefan and I met on our bikes while we were both looking for the same tavern near the Vienna Woods. The place was famous for its fried chicken. We ended up having a few pints together, and the rest is history."

She slathered pâté on a toast. "I think you and I both associate the bike with some of our happiest moments in life. That should keep us fit for as long as our memories last."

"And since we keep making memories on the bike, we should be fit for a good long time." He cut a piece of the duck breast.

"Would you take me to Austria and Hungary to ride?" She didn't think before she asked the question; it just came out.

He didn't miss a beat. "I've already been thinking about that. If everything goes well with the restaurant and bar, by June I may be able to pass the baton to Elian, Luiz, and Luca for three weeks while we travel. The question is, would you rather have a traditional trip—no bikes, but lots of hiking, eating, and drinking—or a cycling-centered trip, where our focus is on covering a stretch of road each day?"

She didn't even have to weigh the two. "The latter sounds more our speed. We're both driven and dedicated. I think we'd get bored with the more leisurely option."

He reached across the table, covering her cold fingers with his warm hand. "I felt pretty sure you'd say that. I already did some research and found out what would be involved in transporting our bikes and bags, securing accommodations, and so forth. I'll show you next time we're at the computer."

She could hardly believe he'd just proposed they spend three weeks on a cycling trip in Europe this upcoming summer. "I'll never live for the next six months."

He gave her his quintessentially Thomas smile: the corner of one side of his mouth tilted up, his eyebrow angled jauntily. "Well, that would kind of defeat the purpose of our trip, don't you think?"

A little later, he ordered the chocolate-pistachio tres leches cake for dessert. It was a toss-up between that and the fried apple pies. But, as she pointed out, one could more easily imagine the taste of the pies than of the cake.

"I like your scientific approach to ordering." Once more his hands surrounded hers across the table. "It's especially useful when there are so many good options."

"It comes from many years of extreme prodigality alternating with extreme frugality."

"Can I interest you in sleeping at my place tonight?" he said coaxingly. "I promise you that wasn't on my mind when I picked you up at the boot camp."

"I'm disappointed, honestly." She pretended to pull a face. "If that wasn't on your mind, what was?"

"What I'd be making us for breakfast tomorrow," he said with a roguish grin.

She pushed his hands away in mock rejection. "You are *too* self-assured, Thomas Gellert." She tried in vain to keep from matching his grin. "What if I say no to tonight?"

"Then I would take that as a sure sign that I've gotten to you."

"How so?"

"Because you'd be doing it deliberately to spite me, and a woman only spites a man when he's gotten to her."

"See? There you go typecasting men and women as you do," she chided, not minding in the least, but playing along for effect. "Suppose I have more man in me than woman?"

"Then I must have more woman in me than man, because what you have fits me to a tee."

"I rather like that image." She sipped her wine. "It reminds me of Aristophanes in *The Symposium.*"

"I suppose the idea of complementary sexes never really goes away after we read that dialogue," he reflected as their waiter placed the cake before them.

"What's your favorite Platonic dialogue?" She took a bite of the cake, closing her eyes in ecstasy at its smoothness.

"The *Phaedrus*," he answered without hesitation. "And you should love that dialogue as well. For in it, you'll remember, Socrates claims that the greatest blessings come through madness and that madness is heaven-sent."

She smiled. "I remember that Socrates punned on the word 'manic' and said that modern men have insensitively added a 't' and called it 'mantic,' which wrongfully sanitizes the idea of divinely inspired mania."

"I'd forgotten that part." He took a bite of cake and finished his wine. "Perhaps most professions nowadays are sanitized versions of the original mania."

"What would your profession have originally been?"

"A shaman," he said with another sideways smile. "We heal through the divine means of alcohol."

"I suppose mine would've been exactly what Socrates did—a sort of peripatetic teaching through dialogue."

"We've kept admirably close to our roots, you and I. Neither alcohol nor teaching has gone out of fashion over the millennia."

"It's true," she said thoughtfully. "We both deal in the universal interests of humankind."

"And now,"—this time his smile was devilish—"shall we go back to my place and engage in another universal interest of humankind?"

She shook her head. "I begin to think you're a man with a one-track mind."

"Only one track matters." Thomas signed the bill with a flourish. "Don't believe anyone who tells you different."

She rode her e-bike behind him to his house. When she'd put the bike in his garage, he said, "Shall we take a shower together?"

Delighted as she was, she cocked her head. "I warn you, I'm pretty sweaty."

"So much the better."

In the bathroom, he peeled off one layer at a time of Nan's clothes, smelling each piece of sweaty clothing intently before laying it aside on the counter, and holding her gaze as he did so. When he had stripped her naked, she advanced towards him and slowly did the same to him, removing one piece at a time and inhaling its aroma before setting it aside. He turned the shower on, and led her in, standing directly behind her, reaching for the washcloth, and soaping it up lingeringly, with his arms encircling her chest. As the water cascaded over them, the smooth slipperiness of their bodies, their proximity, the shapely fit of them together, and the all-encompassing warmth made her close her eyes and lean back into Thomas. She felt the hardened part of him, and her own arousal craved it as she had often sought cold water on a hot day or the fire in subzero temperatures. She brought him into her and the two moved in tandem, her hands on his hips and his divided between her groin and belly. It wasn't long before both erupted, but they remained as one for a while, Thomas continuing to skate his hands over her breasts and kiss her collarbone fervently. That touched off a series of satellite bursts for Nan, which caused him to chuckle in her ear.

"I love hearing you cry out like that," he murmured. "That may be the kind of music I best understand."

Afterwards, when they had toweled one another off, she said, "When we cycle through Austria and Hungary, can we make sure we always have hot showers to return to at the end of each day's ride?"

He nodded. "I couldn't do it any other way. I understand the urge to do bike packing trips and camp out under the stars . . . but I could never do that myself. It seems to punish the body a little *too* much."

She was relieved. "I think of camping as boot camp for the soul. And my soul isn't strong enough to take it."

Chapter Fourteen

The next day, Thomas made them a Dutch Baby—a fluffy popover-like pancake baked in a cast-iron pan over which they then squeezed lemon juice and sifted powdered sugar. With it, he served chicken sausage links and sliced pears.

"You call this a Dutch Baby." Nan savored the airy lightness of the pancake. "But my mom used to call it David Eyre's pancake. It was one of my favorite breakfasts."

"Breakfast is hands down the best meal of the day, in my opinion." He cut up a sausage link. "You can do anything with it, and your hunger makes it taste all the better."

Soon after they'd cleared away the breakfast dishes, she had to leave. "I really need to put in about four hours of solid piano practice. I could feel myself getting just a little rusty when I played with Albert yesterday."

"I'll meet you at your place tomorrow at 9:30 and we can ride up to meet Carl and Len at the reservoir."

"Perfect." She kissed him, gathered her backpack, and took off on the e-bike.

Playing four hours of piano that day wasn't as simple a matter as she'd made it sound to Thomas. For one, she was feeling tired and worn out, and this sort of intensely focused practice required a sharp brain and a well-rested body. Two hours into her mediocre practice session, she decided to try to take a nap downstairs on the bed. But as she'd found over the past year of sleeping more regularly at night, she could no longer fall asleep in the middle of the day. Frustrated after half an hour of trying, she determined to take a run up at the track. After running a sluggish ten miles, she showered, had a snack, and got ready for her student Neal's arrival.

After Charlee's lesson, she had a brief space in which to jot down some ideas for her composition, before riding over to the retirement home to play. She brought with her a watercolor she'd painted in Brooklyn of the Prospect Park lake, as she thought this would fit well on the wall in the common room.

After Nan had finished playing for the residents, Fannie approached her. "Is this for us? Oh, it's lovely." Fannie found a spot on the wall where they could pin it up.

Nan proposed her idea of teaching an informal course in music appreciation and a course in western classical music history in the New Year. "Would this be something the residents would enjoy?"

"I know they would!" Fannie enthused. "Would this same Friday evening slot still work for you?"

Nan nodded. "And could we open the courses up to members of the community as well? Basically advertise on the home's website, put up flyers, spread it by word of mouth?"

"I'll have to find out about that," said Fannie thoughtfully. "Given this pandemic situation, we might have to cap each course at a certain number of outside guests. We'll have to see how many of our own residents show interest."

By the time Nan got home, she was beyond exhausted. She was beginning to feel the pressures mount. She had to be sharp for the long, hard bike ride tomorrow—she didn't want to embarrass Thomas in front of his friends by slowing everyone down. She had to be even sharper for the trio in the evening. And all day Sunday she needed to sustain at least some level of sharpness for her students. She decided to skip dinner and go straight to bed. It was only just past 8, but she might be able to get twelve hours of sleep, which, she felt sure, would make all the difference for the next few days.

She texted Thomas, "Heading to bed. Hope the bar gives you no headaches tonight." Sending off the message, she felt singularly unoriginal. She'd reached the end of her creative and social tether.

His reply came in as she brushed her teeth. "Get lots of sleep. See you tomorrow."

As soon as Nan hit the sheets, she felt the heavy weight of all her limbs, and couldn't even record figures in her journal before drifting off. She turned off the light and fell into a dead sleep that lasted until her alarm woke her at 8:30.

She felt good, but, paradoxically, more tired than ever. It was as if her body had caught on to the fact that all this time she could've been giving it more sleep, and was lodging a complaint against her in response. Long gone by now were her hopes of awaking at 7:30 every morning to compose. She had drifted into that catch-as-catch-can mode of sleeping that she'd had in high school and college. Nan vowed not to let the rotating sleep schedule take over again. She must vigorously re-establish the 10:30 p.m. bedtime and not let anything get in the way of it.

Thomas looked a bit tired as well when they greeted each other. It seemed that burning the candle at both ends didn't suit either of them. They climbed Spruce to the reservoir in companionable silence. At the top, they caught sight of two extremely fit men in their early forties, one about Nan's height and the other a little taller than Thomas. As they pulled up to them, the shorter man opened his arms wide and advanced towards them with a broad smile.

"Tommy boy!" The man enfolded Thomas in a warm, one-armed hug, bumping chests with him while gripping hands. "It's been too long, bro. How you been?"

"Not too bad," said Thomas. "Fatherhood agrees with you, Len."

Len shook his head. "I'm telling you, man, babies will suck the life out of you. I finally decided I had to draw the line somewhere. You gotta have a piece of something for yourself, you know?"

Thomas laughed. "Len, this is Nan. Nan, meet Len and Carl. They set the bar high when it comes to cycling."

Len winked at them. "And low when it comes to everything else. So you guys wanna hit the Three Bears?"

The four of them rolled along Wildcat Canyon, Len chatting affably the entire time.

"Oh, Laura said to tell you hi. She wants to have you over some time soon. We're just starting to emerge from our year-long hibernation."

"I see that," said Thomas. "You doing any racing in the spring?"

"Yeah, a couple crits to start with. Maybe some road races down the coast. You gonna join us?"

Thomas shook his head. "Not this season. The restaurant's going to take all my extra time for the next few months."

Len nodded. "That's big, man. Congrats on getting that going at last. You gonna keep Benny, or replace him with a fancy host?"

Thomas smiled. "Benny remains a fixture at my place for as long as it stays running."

"Yeah, you guys have a mafia vibe going on there." Len spoke admiringly. "No one messes with Benny. I gotta come over there again one of these nights, if I can get free—before you change it up and lose the classic look."

"The new place should be more family-friendly. You can bring Laura and Katie."

Len looked skeptical. "I don't see that happening for a while. Laura's got this magnetic pull to the couch when she comes home most nights. We've ordered in eighty percent of this past year. 'Cause you *know* I don't cook."

Carl interjected. "You know how to make super-dry hard-boiled eggs."

"See? Anybody who can mess up a hard-boiled egg really sucks at cooking," said Len. "So how's Luiz doing these days? Last I texted with him, he was going back to Rio for the holidays."

"He leaves when they start doing the construction and doesn't come back till when I do, after Christmas." Thomas took a swig from his water bottle.

"Lucky bum," Len muttered. "I wish I was going to Rio. Luiz said he's hooking up with three girls down there. He's got the beaches, the mountains, the rainforest . . . and it's summer on top of it all. That's the life."

"Did he also tell you he has about fifty family members he'll have to catch up with?" Thomas grinned.

"See, that's always the problem. And they'll all be expecting gifts, so he'll probably have to bring two suitcases. Then the inspections people at the airport will bust open his suitcases and put fake-apology notes all over them saying sorry they had to inspect his bags, and his suitcases will be ruined by the time he gets them at baggage claim."

Carl embroidered on the scene. "And he'll be jet-lagged, but his mom and aunts won't let him sleep for another sixteen hours while they heap big meals on him and ask him all about his life here."

Len warmed to the subject. "Then all his cousins will arrive, and they'll get pissed when he doesn't remember their names because it's been so long. He doesn't even recognize half of them because they were only babies when he left. And he won't get to leave the house for the first few days because his dad and uncles want to get drunk with him, smoke cigars, and tell really long stories with no point. Poor Luiz."

Thomas laughed. "I'm glad to see Luiz has all of us to cheer him up during his sudden turn of bad luck."

"Tommy, man, you've kept in good shape," Len commented. "The pandemic hasn't slowed you down a bit."

"I could say the same for you," said Thomas. "You claimed you'd lost a lot of your fitness, but you handle these hills just fine."

"It's Laura's Peloton." Len reddened slightly. "Okay, I know I used to badmouth the thing. And you really don't need all those extra bells and whistles. But on some of the days this past year she and I were fighting over who got to use it first. I'm telling you, bro, you start to get really competitive with the other people. And then you want to beat your own personal records."

Carl shook his head. "You'll never get me on one of those. Give me the outdoors any day—rain or shine."

Len countered, "Yeah, but when it's cold and dark after you get home from work on a weekday, what else you gonna do?"

"Dawn patrol's the way to go," Carl argued. "Out on the bike at 6 and back by 8. Into the office by 9."

"The only way you can get me up that early is for a race." Len unzipped his vest. "Then my adrenaline keeps me half-awake all the night before."

"How's the mortgage business going?" Thomas asked.

"You know, same old same old. I'm just glad I'm an independent broker these days. The guys I used to work within the bank are having to initiate foreclosures on hundreds of homes. Those Covid mortgage bailout programs just expired, and they left a lot of people high and dry."

"I suppose their best option at this point is just to sell their homes and make whatever profit they can?" Thomas suggested.

Len nodded. "Which is great for the housing market, of course. There's been a short supply and high demand ever since the pandemic. In just a year, home prices are up twenty percent."

"So who are most of the people you're dealing with these days?" Thomas took out a Power bar, opened it, and tore off a chunk.

"Most of the time, you gotta be well-off to buy in this market out here in the Bay Area," said Len. "You need enough to bid way above the asking price, you gotta have lots of cash, perfect credit, and good savings. Personally, I enjoy working with the middle folks, the guys who need someone tough in their camp to negotiate a good loan in the face of stiff competition."

Thomas grinned. "I'll bet you're really good at that."

Len shrugged. "I wouldn't have gone independent if I wasn't."

Nan listened to them talk, glad to have the chance to see Thomas in another element, among his friends. As they climbed the ever-challenging "Pig Farm"—a steep hill on Alhambra Valley Road in Briones—a huge truck gunned by them at high speed, nearly clipping them all, even as they hugged the right-hand side of the road, and even as the left-hand side of the road was empty.

"Fuckin' asshole!" shouted Len, taking his hands off the handlebars to give the driver both of his middle fingers. "You always get at least one moron on this stretch of road. There should be a sign on the side of the road back there that says, 'Warning: assholes ahead.'"

"That would probably only encourage them." Thomas chuckled.

"Yeah, once an asshole, always an asshole," Len groused. "You can always tell from the motor what kind of driver you're going to get. I could hear that guy was in a souped-up Tacoma from his engine. That tells you everything you need to know."

Carl said, "The worst is when they have a trailer in back and you realize it at the very last minute, and there's nowhere else to go on the right."

"In that situation, it's best not to know a trailer is coming," said Len matter-of-factly. "You might freak out and steer into it."

They climbed Reliez Valley into Lafayette and stopped off at a school to get some water. Len observed, "Your girl here is pretty fast, Tommy. Now I see how you've kept so fit."

Nan laughed. "I think it worked the other way, actually."

"I'm jealous you guys get to ride Wednesdays too." Len ripped off the top of a GU shot and downed its contents in one smooth motion. "That second long ride makes all the difference in your training."

"In that sense, it's nice to own a bar," Thomas admitted.

"Man, if I owned a bar, I'd train in the middle of every day, I'd sponsor a team, and I'd close for at least three months every summer to ride through France and Spain."

"I can see you'd throw away a lot of money." Thomas winked at Nan. "Maybe it's a good thing you're in the mortgage business."

"Yeah, I don't think I could resist all that booze either," Len conceded.

"You get to the point where you see so many other people drinking, you stop wanting to do it yourself," said Thomas.

"That's news to me," Nan couldn't help interposing. "You keep up a pretty steady pace in the drinking department."

"Ahh!" Len burst out laughing. "See, Tommy boy. She keeps you honest."

They climbed over Moraga Road and up Pinehurst, with Len and Carl saying their goodbyes and heading down Shepherd Canyon, and Thomas and Nan continuing north to descend Claremont.

"Nice guys, huh?" said Thomas, as they rode along Skyline Boulevard.

"Yeah. I noticed Len never talked about his baby, though. Does he like being a dad?"

"Oh, yeah. Don't let him fool you. You should see him with Katie—he's in love. He just doesn't like to wear it on his sleeve."

"You're right—he does ride like Peter Sagan. Aggressive, with a sprinter's confidence and really bold moves. I'll bet he wins a lot of races."

Thomas nodded. "He does especially well in crits. I used to race with him back before the pandemic."

"I imagine in racing, you're better at the longer road races?" she asked.

"You know it, from riding with me this past year."

"Do you miss racing?"

"A little, sometimes. But I think I've moved on to other kinds of cycling now. Touring is more my speed these days."

"Then you get to eat, drink, and talk along the way," said Nan, a sparkle in her eye.

"Precisely."

♩ ♩ ♩ ♩

Later that afternoon, when Nan arrived at Jules's house, Simon was miccing up the performance area, Jules's wife Isabel was laying out dishes and glassware on a table, Simon's son Jonathan was flirting with Cécile, Céline and Arthur were enlisting the help of Jules's son Josh and their own son Max to unload a bunch of chairs from their truck, and Jules was lighting two heating lamps out on the deck—one his own, and one borrowed from a neighbor.

After joining her bottles of wine to the collection, Nan helped Isabel set things up at the table. Admittedly, the event was more of an excuse for people to gather together with music as the premise, for the trio only took twenty-eight minutes to perform. But Nan had been pushing the members of the group for some time to add on works for solo strings and piano in order to flesh out the programs of their concerts. Now that everyone in the quintet had okayed the idea of putting their material up on a YouTube channel, Nan felt they could really start to soar, both in recordings and performances. She also hoped that all these recordings, performances, and her completed composition would make her CV much more substantial and impressive. She even looked forward to a day when the quintet would perform her own quintet.

Lilian and Lewis arrived first, and Nan welcomed them warmly, introducing them to everyone. Then Bess and Bertie came, Lilian expressing her delight to see how Bertie had changed since she'd last seen him. Among all three of the Arbuthnot siblings, Lilian had a particular fondness for Bess, whom she'd taught first. Lilian, Lewis, Bess, and Bertie took seats inside, as more people gradually began to filter in. Jules's eighteen-year-old daughter Paulina brought her and Josh's cousin Stacy back from the airport next—Stacy was visiting from Cincinnati. Next the ever-effusive Despina bustled in with a large entourage of family members who took food and drinks to the kitchen and dining room while Despina engaged Céline and Nan in conversation. Despina's youngest, Jason, who was eight, made Bertie's night by sitting next to him to play his video games, asking him which games he liked. Paul, his wife June, and their three kids came next.

"Jules, I gather you're to be relieved of your duties come March," said Paul in a rich baritone voice.

"It's been great for me, never having to move a foot from home when we meet," said Jules genially.

Paul deposited a large plate of cheeses wrapped in plastic on the table. "I hope we won't have to put the food and drink on pause when we perform at Gregorio's. That's a critical motivation to our playing."

Despina came over. "Drinks yes; food, maybe not. He has a front room where we can set up some basic wine and cups. But the sound is more than going to make up for it."

At last Margot and Thomas arrived. Nan tried to curb her elation at seeing both of them. It felt like such a strange juxtaposition of contexts, to have her chamber group, Margot, Thomas, Bess, and Bertie all in one room together at once. Margot and Thomas took seats on the deck by one of the heating lamps. Nan felt thoroughly supported and knew this performance of the "Ghost" was going to be her best.

After everyone had taken their seats, the musicians tuned their instruments. Jules stood and spoke a bit about Beethoven's middle period, his friendship with the Hungarian Countess Marie von Erdödy that enabled the writing of this trio, and Czerny's nicknaming of the piece. Nan heard Bertie's small voice asking Bess if there was going to be a real ghost in the room. Nan smiled, and then, at last, the three musicians launched into the first movement.

As they played the tense and ominous second movement, Nan was in seventh heaven. It felt not only like the culmination of so many hours of thought and work, but a new piece because of the people who'd gathered to hear it. Jules and Céline rounded off the movement in quiet pizzicato unison with her.

Then followed the sprightly third movement that so abounded in dancing and lively conversations among the three instruments. Nan was especially grateful to have Cécile turning pages for her in this movement, which afforded the fewest chances of a respite for the pianist. The trio rounded off the piece with the sort of enthusiasm and flair that comes only with a live performance. They smiled broadly at one another as the audience clapped, and they stood, bowing their heads.

"Bravo!" Paul boomed sonorously.

Nan saw Lilian beaming at her, Bess and Bertie applauding heartily, and, over on the deck, Margot and Thomas smiling unreservedly as they clapped. The moment felt pregnant with good will and common purpose. Nan hoped that her students would share this same moment when they gave their own recitals over the next two Saturdays. Then everyone began to relax, move, and talk, and the focus shifted to conversation, food, and drink.

Nan helped Lilian, Lewis, and Bess to some wine, and Isabel gave Bertie a fruit soda drink. They spoke of a cooking show that Lillian, Lewis, and Isabel all loved watching, and the charms of the show's host and his family. Thomas appeared to have been waylaid by Despina and her husband, who had caused him to throw his head back in laughter at something they'd said. Margot wended her way towards Nan, wrapping her in a close hug.

"That was *such* a lovely performance," said Margot. "You all are incredibly talented. I see why this is your favorite Beethoven trio. It's got so many contrasting moods."

"Remember when we heard the 'Archduke' trio at the Fogg Museum?" Nan asked, recalling a concert about eight years back in Cambridge.

Margot nodded. "That was sensational. Maybe this is becoming a theme for us—working our way through the Beethoven trios in concert, together."

"We should go to a concert in the City again sometime soon. It's been way too long."

Paul came over and clapped a hand on Nan's back jovially. "Wonderful work, Nan. I look forward to playing our Schumann in the New Year."

"And do start thinking about what you might want to play solo or with piano," Nan urged. "We can fit in another two to three pieces in addition to the quintet."

"We'll see," said Paul, who played viola. "I may be too busy to do a solo this time around."

Nan could feel Thomas's presence before she saw or heard him. It was as if a large wave was about to crash on shore, and she embraced the wave's impending force.

His warm voice sounded at her side, causing her knees to weaken. "That was really good. You were all conversing in dialogue with each other."

"That's why I love chamber music so much." Nan once more felt the rush of confidence that came after a good performance. "Well, that reason and the fact that it's designed for a more intimate setting than orchestral works."

Isabel handed Thomas and Nan glasses of wine and then disappeared into the kitchen to bring in more plates of food. Bess came over, and Nan introduced Thomas to her.

"I hear you're taking a huge leap and opening a restaurant in your bar," said Bess, sipping her wine. "You must be nervous, with the new Omicron variant and all."

"I subscribe to the theory that Covid will soon become just like the flu, with new variants each year and everyone getting their updated shots at periodic intervals," said Thomas. "So far, most of the detected Omicron cases in the U.S. haven't been very serious. And they still don't know much about its rate of spread."

Bess shook her head. "Scientists have said it has more than thirty changes to its spike protein. That's enough to make us all worry."

"I think this may be a case of a little information being a dangerous thing." Thomas took a sip of wine. "I'm not equipped to be an amateur epidemiologist or virologist. I'll wait until the public health officials tell us what the bottom line is for policy in dealing with the new variant."

"I disagree. This is one case where the public can't afford not to know every last discovery in science," Bess countered. "That's like accepting when a few top economists reassure us the economy is doing well, and then a financial crisis happens and topples everything."

"I would argue the two situations are very different," said Thomas, clearly enjoying measuring swords with Bess. "It's much easier for individuals to acquire a basic knowledge of economics than for them to become scientists. If I were a scientist, I would feel insulted if the entire world tried to be casual specialists in what I had methodically devoted my training and research to for decades."

"Then why do those same scientists so willingly share their latest findings with the public?" asked Bess provokingly.

Thomas smiled. "Ego, for the most part. Everyone wants to be in the news nowadays, and they're all jumping at the chance to become famous, if even just for half-baked theories that haven't yet been fully proven. I think the scientists you mention are irresponsible for scaring the public before Fauci and his aides can shape a sensible policy out of their data."

"So it's a case of questionable motives?" said Bess. "I guess I've seen my fair share of those, as a stats person."

Céline came over, taking Nan's attention away from Bess and Thomas's continuing conversation. "What do you think of the Debussy cello sonata for the March concert?"

Nan's eyes widened. "Oh, Céline, I've always wanted to play that! Let's do it!"

"I need something to take my mind off the fact that Cécile will be leaving us soon." Céline lowered her voice conspiratorially, and leaned closer. "Nan, you have a very handsome fellow there. Does he play as well?"

"He plays other things—but not instruments." Nan matched her smile.

"Next time, we'll have him to dinner with you," Céline declared. "He seems very sharp."

Guests continued to shift about like tides, with unforeseen combinations of personality and interest forming from one moment to the next around the room. As Nan made up a plate of food for Bertie, she overheard Lewis, Thomas, and June speculating about why it is that urban centers in America are so much less dense than those in Europe. Lewis pinpointed crime as a major reason Americans flee their cities for the suburbs, and Thomas blamed America's long-standing investment in and subsidy of cars. June noted that Europe made public housing attractive and available to middle-income residents, so that people from a wider range of economic backgrounds could live together in the cities. Nan was eavesdropping avidly when Bess came over.

"Remember, Bertie doesn't like anything too sour—he won't like those pickles," said Bess. In true Bess fashion, she segued naturally from the pragmatic to the aesthetic. "That was really beautiful, Nan. Your group seems to have grown together nicely—as if when you play now you're more like one instrument with three voices."

Nan was deeply touched by these words. "Thank you."

"Now you've got one more recording and performance for your CV. How's the composing going?"

"I've nearly finished the first movement and am about halfway through the third. The second movement is really tough, but I'm plugging away at it whenever I get the chance."

"You're still aiming to finish by December 31?"

Nan nodded. "I wish I could work on it twenty-four seven, to be honest."

"If you have that kind of approach, you really will finish it by the end of the year," Bess remarked knowingly. "That'll be one more thing for us to celebrate in Minneapolis."

As Bess took Bertie the plate Nan had prepared for him, Paulina, Stacy, and Cécile came over to Nan. Nan got the distinct feeling that Paulina and Stacy were a bit in awe of Cécile, who seemed to be the confident leader of the group. Cécile said, "Nan, I wondered if you might be able to use a vocalist for a few pieces in the March concert."

Nan beamed. "What do you have in mind to sing?"

"Some Grieg, and, if it wouldn't be too much Debussy on the program, some Debussy songs as well?"

"You can never have enough Debussy," said Nan decisively. "Let's go for it."

Nan didn't add that Debussy had been her mom's favorite composer, and that she was always glad to resurrect Gwen by playing Debussy.

By the end of the evening, Nan had convinced Despina to return to a piece she'd played in college, Brahms's violin sonata number one in G. They now had a complete ninety minutes of music programmed for their March concert, and Nan brimmed with excitement. When everyone had said their goodbyes, and as she, Margot, and Thomas walked out to his car, Nan couldn't contain her buoyancy, spinning around several times with glee.

Margot cried delightedly, "You look like a dancing top!"

Thomas said, "Someone needs to weight her down."

Margot winked at him. "That's your province."

♩ ♩ ♩ ♩

Since it was only just past 9, they decided to head to Thomas's bar, Nan riding her e-bike and following Thomas and Margot in his car.

Nan couldn't help comparing this bar visit to their visit a month ago, when she'd been so nervous she couldn't read the menu. Thomas didn't bother to ask them what they wanted this time. He just set two cocktails in front of them that probably weren't even on the menu. His only accompanying words were "Inspired by what you ordered last time."

"Won't you join us?" asked Margot.

"The time that two friends who are separated by the Bay spend together is very precious," said Thomas with a quiet smile.

When he'd left to help Luiz out, Margot sighed. "Now, *that's* a boyfriend."

"What's going on with Aaron?"

"I called and texted him yesterday to see how the interview went. No response. He didn't suggest anything for this weekend, and here we are on Saturday night, me over in his neck of the woods and no Aaron." Margot took a forlorn sip of her cocktail. She couldn't help humming over its deliciousness.

Nan placed a hand over Margot's. "Wait just another day or so. I'm sure he'll call or text. He's probably keyed up about Google's decision."

"You know what? I'm sort of tired of having all of this be about him," said Margot. "If he doesn't call or text by the end of the weekend, I'm going to suggest we take a break. He's got to give me something to work with."

Nan nodded. "If that's what you have to do, then do it. Ultimately, it's about you remaining sane."

Margot looked around briefly before she spoke, and cracked a roguish smile. "If I remember correctly, that isn't exactly the tack you took with Thomas."

Nan burst out laughing. "I don't think I 'took any tack' with Thomas. He was more like a whirlwind that swooped me up in its path."

"That's the way it's supposed to be." Margot spoke with wistful romanticism. "In a relationship, you're supposed to feel as if you're sucking the marrow out of life—to switch metaphors."

Nan became serious. "But that's sort of the problem. My mind is beginning to race a little too much again. Apart from Friday night, I haven't been sleeping very much, and I find myself constantly thinking about either Thomas or the composition. I feel as if even the small bouts of sleep I get are light and fitful because my brain is still turned on. I want so badly to sleep, but I don't ever want to *be* asleep, if that makes any sense."

Margot looked worried. "Can you spend a full hour before going to bed just tuning out and reading or playing piano or some other relaxing activity to get your mind off of these things?"

"I always mean to, and then I break my vow."

"If it helps, you can hold yourself accountable to me. Text me when you're starting your hour-long period of relaxation, and report the time you plan to go to bed. Then immediately put away your phone and laptop."

"I am so lucky to have you, Margot." Nan squeezed her hand gently. "Text or call me any time tomorrow about Aaron."

Margot left around 10:30, and Nan had lost herself in a reverie when she felt Thomas's incomparable heat behind her. "Would you like another drink?"

Her entire body thrilled, as always, to have him so near her. Would this feeling ever go away? she idly wondered.

"I think I'm done for the night, thanks."

"I can get out of here in five minutes if you'd like to come over."

She knew, in her heart of hearts, that she ought to say no and go back to sleep a solid night in her own bed. But Thomas's smell a mere few inches from her, and his smooth voice, together with his electrifying presence, decided the matter before she could speak. He seemed to read from her expression the answer to his question, and he smiled and got ready to leave.

They stopped off at Nan's place so she could collect her night things and, most importantly, take her pills. She dropped off her e-bike, and rode with Thomas in his car. On the way, he looked over at her several times without speaking.

"What is it?" she asked.

"You look really tired. Are you okay?"

"Now I am." She smiled. "You know I love being with you."

"Let's take it a little easy tonight. Maybe focus more on sleep."

"That's what we planned to do when the ants came, remember? And look where that got us."

"True," he said. "Why don't we set a new challenge for ourselves tonight. See who can resist the other's charms the longest. That ought to appeal to our deep-seated competitiveness *and* allow us both plenty of sleep."

She was skeptical. "In theory, your plan might work well. But in practice, unconsciously in our sleep one of us initiates something and the other responds—who's responsible for what follows? Or supposing our mere proximity starts us both off simultaneously . . ."

"I have a king-sized bed," he reminded her. "I have faith we can keep our distance enough to fulfill the challenge."

"What does the winner get?" An intrigued glint entered her eye despite herself.

"A full-body massage from the loser tomorrow morning," he said with an even more devilish glint.

"Deal," she said.

As he took off her coat in his living room, he kissed the base of her neck seductively, ending by sucking for two seconds on a spot above her collarbone and blowing cool air on it.

She moaned, and leaned back into him, before crying, "No fair! If you're going to seduce me, then I don't stand a chance tonight."

"Seduce me back then," he said equably.

"All right, you asked for it. Now remember, we don't get to have sex until after the massage tomorrow morning—and the loser has to give a full-body massage to the winner."

Looking as if he was fully in his element, he nodded, awaiting her next move.

Knowing a few of his weaknesses by now, she led him over to the couch, gestured for him to take a seat, and sat on his lap. She gently closed his eyelids, placing one of her palms over his eyes, and with the other hand, she ran her index finger over his lips, coaxing them to open up to her. She let him hold on to her finger deeply, before slowly removing it and placing her middle finger gradually between his lips. She followed this with her ring finger, allowing him to suck on it for a minute before withdrawing it. She then settled a bit more comfortably onto his groin and traced the contours of his face, jawbone, and Adam's apple with her fingers, still keeping her other hand over his eyes. She undid a few of the buttons of his shirt and continued to run her fingers slowly over his pectorals.

"Oh, those pianist's fingers!" he groaned. "So delicate and so full of delightful torture."

She felt his rising bulge beneath her crotch, and began to climb off of his lap.

"Not so fast." He pulled her back. "You don't get off that easily. It's my turn now."

She was surprised at the alacrity with which he shifted gears and resumed the role of seducer. "But how far do we go with this? You know what happened last time."

"Just a little further, and then we'll stop." He drew her hair to the side, placed his tongue in her ear, and swirled it around infuriatingly while cupping her neck in his free hand. She fought in vain to resist this triple onslaught of tongue, breath, and hand: however much she balled up her fists and pressed them hard into his thighs, her mind was crowded with all-consuming images of Thomas behind her, Thomas on top of her, Thomas inside of her. He tugged gently but assertively at her hair and moved his free hand to her breast, while planting a few teasing kisses on the nape of her neck. Then it was his turn to shift in such a way as to drive her insane as she felt his hardness at her points of highest sensitivity. As if surmising her weakness, he grazed her groin with his hand and then skated his fingers along her inner thigh.

"Ohh . . . you are the world's biggest cheater," she sighed. Then, just as she found herself leaning into Thomas and allowing herself to fall completely within his alluring ambit, she pulled herself up short. Remembering the glory and the prize to be won, she reached into her innermost reserves of discipline, and turned her torso so that she was facing him. His eyes met hers and seemed to read in them her fierce determination. Rather than testing her any further, he withdrew his hands and placed them around her hips. He lifted her onto the couch beside him, and adjusted his pants.

"I always mean to be a gentleman around you, but lately my best-laid plans have gone awry."

"A gentleman!" She laughed in disbelief. "A rake, more like."

He had the grace to look abashed. "I warned you on our first date my motives were self-interested."

She became more serious. "It pleases you to claim as much, but that too is another role."

"Though no less real for all that."

"No, I'll grant you that. But insofar as it's only part of the whole picture, it isn't true."

"Well, the gentlemanly part of me genuinely wants to see you get some sleep tonight. Shall we turn in?"

"Thank you, Dr. Jekyll. I'll take you up on that suggestion."

"Remind me to punish you tomorrow morning, however," he said enticingly, drawing her back in for one more kiss.

"For what?"

"For being so charming."

She lay like a zombie that night in his bed. If he was tempted to take advantage of her nearness, she never detected it. Both their bodies seemed to recognize that a truce had been called: the two of them slept like rocks, and only around 9, when the sun was finally starting to break through the cloud cover, did they come to life once more. He turned his body so that it was flush with hers and, placing his hand on her belly, said, "I think we've both earned a full-body massage."

"But who gets one first?" she asked drowsily.

"You do," he said. "You exerted more considerable restraint last night."

In the event, they didn't get beyond Nan's massage, for greater stirrings prompted them.

Chapter Fifteen

Nan was apologetic for making Thomas drive once more to her place at an off hour. As he pulled up in front of her apartment at 10:45 that Sunday morning, she said, "I'll make it up to you for having to shuttle me back and forth so much this week."

Thomas shrugged, smiling. "It's the best use of this car that I can think of. I hope you can get some serious sleep tonight after you've finished teaching." He examined her eyes with concern, much the way her mom used to do when she was determining if Nan was too sick to go to school. "And I hope I haven't derailed you too much."

"Uh-oh." Nan feigned alarm. "If it's that obvious I'm missing sleep, I must look pretty bad. Don't worry, though. If I hadn't been *with* you, I would've been thinking *about* you, and that wouldn't have been great either."

"Much as it's nice to have you thinking about me, I'd rather have you sleeping. Let me know what I can do to help this week."

"I will. Thanks." They kissed, and Nan stepped out of the car, watching for the third time that week as Thomas did a swift and confident one-point U-turn on College, waving to her as he sped off.

That evening after Nan's last lesson, Margot called. "Aaron got the Google position. But I told him I wanted to take a break. I told him he seems to have a lot of stuff going on right now and maybe it's not the right time for us to try to develop a relationship. He agreed, but he said he'd like us to be friends. I asked him how that would work if he doesn't have any time, and he said he will eventually have time. I said he seems to be looking for a much lower-pressure relationship than I am. But in the end, I agreed that for whatever it's worth, we could remain friends."

"I'm really sorry, love," said Nan consolingly. "But maybe there's a chance in a couple of months that you guys will be able to pick up again where you left off?"

"Who knows," said Margot resignedly. "But for now, I need to feel I'm getting more out of the relationship than I am, with a little less work."

"I understand. Well, any time you want to talk about it more, I'm here."

Monday afternoon, Nan shared with Naseer her desire to channel the long-term brilliance that Thomas had extolled at Sue and Manuel's party. Since the party, Nan had become increasingly puzzled by the mystery that was the passage of time: how could she mentally manage it in the short term so that she accomplished what she needed to in the long term?

"I wrote down in my journal the three things that matter most to me right now as I'm finishing off the year 2021: Thomas, my composition, and surviving," she said.

"The first two sound comparatively simple." Naseer smiled. "But survival can be a complicated thing."

"Yes," she conceded. "I think that's because it involves both mental and physical well-being. I still don't fully understand how these are connected. I know that sleep is critical. I also know that in order to sleep, I need to be largely free from anxiety, and that means clearing myself of financial worries." She didn't add that she'd begun casting about in earnest for ways to make the money to cover the minimum payments on her credit cards and the utility payments that were coming due.

She told Naseer of how she'd begun putting in practice Margot's idea of texting Margot when she was winding down each evening. He took note of this and asked if it was helping her sleep.

"Yes, only the more engrossed I become in my composing, the more my body insists on waking up around 7:30 a.m., regardless of the fact that, by my own estimation, I now owe over 28 hours of sleep. So while I'm able to recoup a little bit of sleep in the evenings, I can never do so in the mornings."

"Let's celebrate the small stuff, though," said Naseer. "I notice you've managed to focus on composing, sleep, exercise, and your new partner. That's quite an accomplishment."

Nan supposed he was right, and she felt pleased to discover the healthy pattern she'd fallen into. On her way out of his office, she was lost in thought. What if Thomas was right in calling the long seduction "art"? The timing of their lovemaking resembled the timing of creating a work of art: you had to add little bits and pieces gradually to the whole without knowing, in the end, what the whole would amount to, and yet trusting, all the while, that your small efforts would eventually pay off. The frustrations of holding back with Thomas paled in comparison to the frustrations Nan had felt over the past few years of projecting and deferring; but she couldn't deny the close parallels. Now, here she was, falling in love and composing at long last . . . she knew it couldn't be just a coincidence. Falling in love, like creating at its best, allowed her to lose herself in the moment and not overthink how the details related to the big picture.

On Tuesday afternoon, when she was running up towards the track, she caught sight of an old dresser that had been put out on the curb on a side street leading off of Claremont Avenue. She ran to take a closer look at the dresser's condition. Apart from the fact that it was missing a handle on one of its drawers, it was perfectly suitable for holding clothes. While this wasn't the style of dresser she would normally consider for herself, she had an idea: what if she could place her clothes in this dresser and take them out of the antique bureau she had down in her bedroom, which was easily worth $300? She could sell the latter on Craigslist and clear her debts for December.

Nan removed all the drawers of the dresser first. She took two of the drawers back to her apartment and returned with bungee cords and a dolly that her landlord kept in the backyard. She placed the dresser on the dolly, secured it with the bungee cords, and trundled it back to her place. Then, finally, she returned with the dolly for the final three dresser drawers. Nan did a truncated run and, upon returning, began to empty all of her clothes out of the downstairs bureau. Now, how to hoist it up the winding stairs to her living room? She decided to enlist Thomas's help the next day before they headed out on their bike ride.

Wednesday morning, when Thomas arrived and heard the plan, he laughed, but he good-naturedly put his thinking cap on as to how they could get the tall bureau up the relatively narrow eleven winding stairs to Nan's main floor. They found very quickly that tilting it horizontally wouldn't work. Instead, Thomas suggested they keep it as vertical as possible as he took the bottom side and Nan took the top side. They slowly hauled it up, and then, since Nan was determined to have it ready for outdoor photos, they carried it down the thirteen outside steps of the apartment into the back driveway.

"I'll take the photos after we come back, while we still have good afternoon sun," said Nan. "And then I'll cover it with a tarp until someone from Craigslist can come pick it up."

"It's a nice bureau," he remarked. "You should have no problem selling it. Shall we bring the dresser you found down to your bedroom?"

As they rode up the hill, he looked over at her a few times as if he wanted to say something, but was conflicted. Finally, he said, "I know sometimes finances can be tough for you. I hope if you were ever in a tight spot, you wouldn't hesitate to borrow money from me."

She was surprised at his intuition, even as a wave of fondness and gratitude swept over her. "Oh, thank you. That's very generous. But I'm fine for now."

That afternoon, after she'd listed the bureau on Craigslist, Nan packed all of her own clothes into the dresser she'd found. It was a tight squeeze, but she managed it by placing the clothes that she rarely wore in a plastic bin in her closet. She figured as long as she sold the bureau by Saturday, she could beat the rains that were predicted to cover the East Bay from Sunday through Thursday of the following week.

That same afternoon brought a pleasant surprise in the form of an email from the director of the School of Music at San Francisco State University. Nan had completely forgotten that she had applied for a lecturership in the Music B.A. program at SFSU some time back in October. The director wanted to interview with Nan on Friday over Zoom; Nan emailed him back immediately with a few time slots that would work for her.

She called Thomas. As soon as he picked up, she announced, "I got an interview!"

"Congratulations. Where and when?"

She told him all the details with unrestrained ebullience. "I'm not sure I stand a ghost of a chance, but the interview will be a good experience none the less."

"I was going to suggest that we work on a website for you so you can refer people to it when you're making applications."

"That would be wonderful. I don't have the faintest clue how to do that."

"Then you can have links to things like your group's YouTube channel, a performance calendar, audio and video clips, your CV, and so forth," he pointed out. "Would you like to work on it this evening at my place?"

"How can I say no to that? But is it possible for us to be in bed by 10:30?"

"We'll make it a priority. Bring your laptop so you have all the files you need."

Thomas picked her up at 6, and they stopped by Nabolom Pizzeria in Elmwood to get a pizza to take back to his place. By way of explaining this splurge, he said, "So far as I can tell, interviews for academic positions are a big deal."

"Will we need wine to go with it?" she asked.

"I have plenty at home."

She had brought her road bike so she could ride back to the UC Berkeley campus the next day without troubling Thomas.

They ate and drank on the floor of his living room—"just so we don't have to do any more chair-sitting than necessary," he reasoned—before setting to work on Nan's website at his standing desk. They didn't have any trouble securing her preferred domain name, @nanarbuthnot.com, because no one had taken it yet.

"Well, that was easy," he remarked. "That can often be the hardest part."

After registering her domain name, they signed her up for a web hosting account, and pointed her domain to her website. Then, since Nan had a Mac, they used BlueGriffon to write some of the basic content of her home page. Here, he urged her to be fully relaxed in her writing. "You'll come back to all this later and edit it. For now you just want some plain placeholding material."

By the time they had tested her home page in a web browser, he noticed that the time was 10:15 and said, "Shall we call it a night? We can tackle pictures and logos tomorrow."

She placed a hand on his arm. "Thank you. This is really exciting. I'm not sure I'd have been able to do this on my own."

"Well, you would have," he said with his trademark smile. "But you may have just needed that extra push to do it sooner rather than later."

Although they made it to bed before 10:30, Nan was still wired. She reached over and ran her fingers through Thomas's thick hair. "Can I draw your portrait?"

"Now? Aren't you sleepy?"

"Not yet. This will relax me enough to get me to sleep."

"Sure. I don't feel particularly tired yet. What do you want me to do?" He propped himself up on an elbow.

"Let me first grab some paper and a pen."

When she came back, she had him sit up with a pillow behind his back against the headboard of the bed. As they talked about the steps for the construction of the addition to his bar, he naturally resumed the attitude she loved best of all: he gazed off into the middle distance as if his mind was fully absorbed in a problem that took precedence over his immediate surroundings. She drew him in profile from his left, his jaw line covered by a shadow of stubble and his dark eyebrows, as usual, sloping slightly downward towards his delicately arched nose. As she drew, she tried to capture the subtle contours of his cheekbones and temples, the brilliance of his eyes, the expressiveness of his lips, and the breadth of his forehead. Her pen found even his ears and Adam's apple shapely. She held the drawing away from herself for a moment, closed her eyes briefly, and then opened them. After adding a few more highlights, she felt this was as far as she could go tonight.

"Can I see?" he asked.

"It could probably do with some further touches . . ." Nan edged over next to Thomas and showing him the drawing. Privately, she was no little pleased at her work: she found herself inspired in a way she hadn't been for some time now in drawing. Once more—for the first time perhaps since her mom had died—she felt a force greater than herself had guided her hand.

"Wow." He didn't say anything for another minute. He merely examined the portrait. "You really flattered me."

She smiled and shook her head. "No, I just drew what was there."

He looked steadily into her eyes and she felt as if he was boring into her soul. Had he read in the drawing how deeply she cared about him? She found herself blushing.

"You can have it," she said hurriedly. "It's only a first attempt."

"I'd like to frame it and hang it up behind the bar," he said. "Would you sign it for me?"

Nan did so, and then Thomas put out the light. He eased her onto her back and shifted so that he was above her, still holding her gaze. He softly stroked her breasts, hips, and thighs, and she held on to his shoulders like a ballast as she gradually heated up and began to dissolve in his caresses. They kissed, and their one kiss undid Nan entirely. She cried out for Thomas to enter her and he slowly came in, establishing a sporadic rhythm of advances and retreats that drove her wild. Before long, they each felt the mounting climax of the other and this pushed them both over the edge.

Soon after, Nan fell into the soundest sleep she'd had since Saturday night.

∫ ∫ ∫ ∫

The next morning, Nan made Yankee-style cornbread for them, after a recipe from Gwen that she had long ago memorized. They cut square pieces in half, slathered butter on them, and drizzled honey on top. Thomas made bacon and pourover coffees, and they sat at his bar counter enjoying a few moments that made Nan feel like the luckiest woman in the world. Since Monday she had increasingly begun to sense that the combined processes of falling in love and immersing herself in her composition were working a transformation on her. She had discovered that her counting and her inability to live in the present were tied together: she'd been so afraid of losing track of her counting that she'd actively prevented herself from giving herself over completely to the present moment. Now she could tell she was savoring small details again, and really appreciating moments like this one, even as she was letting some of her counting tendencies slide a bit. In addition, her anxiety about pursuing excellence in all fields at all times had partly fallen off. Paradoxically enough, this relaxation had enabled her to create some of the first good poetry and art she'd made in a few years.

Sitting with Thomas over a breakfast Gwen had made many times during her childhood recalled Nan to a moment of deep grief that she had experienced soon after her mom had died. She hesitated for a second before deciding she would open up to Thomas about it.

"This cornbread was one of the breakfast staples my mom always made when I grew up."

Thomas leaned in and met her eyes with an intensely focused look.

"Once, a few days after Mom died, I went on a late-night walk around my neighborhood in Brooklyn. I suddenly dropped down on the sidewalk shuddering with sobs and moaning as I caught sight of a few stars looking down at me from above. For several minutes my unconstrained wails ripped through the silence of the residential street—where not another soul stirred—as I realized my complete aloneness in the world."

Thomas placed his warm hand on Nan's. "At that moment, it seemed as if I would never smile again, much less laugh." She gazed into Thomas's softened eyes. "Now—it feels like a small miracle—for the past few weeks, I've once again been able to see the lighter side of things. I can relish the comic potential that lies in details—like an orca I saw in my pancake one morning, or Bertie asking after the ghost as I began to perform my trio the other night, or Carl insulting Len about his hard-boiled eggs, or the look of feigned offense that Dan Fisher shot you in response to your ribbing him. Life is yielding laughter again.”

Thomas wrapped his free arm around Nan’s back. She leaned towards him and rested her head on his shoulder. Neither of them said a word, for it seemed none was needed. They remained like this for a few moments. Then Thomas broke the silence.

“The closest experience I have to what you describe is losing a good friend from college three years ago.”

Nan heard a strained note in his voice and she angled her head towards his eyes, which had clouded over. “Did he also die of cancer?”

He took so long to answer that she began to think he hadn’t heard her. Then he spoke in a low voice. “He hanged himself.”

Nan placed her arm across Thomas’s waist. She whispered falteringly, “I’m so sorry. What was his name?”

“Yevgeny Ivanov. He was my dormmate for three and a half years. We did almost everything together. After college we even shared a two-bedroom apartment for two years in the City. We played hundreds of hours of chess together, and he kept me sharp in my tactics. He was always trying to make it as a writer and never quite succeeding the way he’d hoped. He kept taking menial jobs at factories and hospitals to support himself, and I lent him a lot of money I never expected to be paid back. I thought he was a really gifted writer. He had a blog for a while that had a few faithful followers who appreciated its rare kind of humor. He always joked that it took a Russian to see the funny side of bleakness and tragedy. But he had a depressive nature and he always drank heavily.” Thomas paused, took a breath, and then continued. “In the end, I think there was too wide a gulf between his high expectations of himself—or the expectations he imagined his family and friends had of him—and what he saw himself accomplishing. Several of his friends, including me, witnessed him spiraling into despair during the last few years. The pieces of writing he sent us should have told us what he was preparing to do, and as his girlfriend said, he talked more and more about the virtues of death and being dead. I’m particularly guilty because as his oldest and closest friend I knew him best of all. I should have seen the handwriting on the wall.”

“But how could you have known what he was going to do—or prevented him from doing it ultimately?” asked Nan gently, pulling him into a firm embrace.

“I let him down. I was the person who knew him the longest and I’d always been aware of the darkness within him. At times, I even suspected he was capable of suicide. He was always fighting so many demons—as even his funniest stories revealed." Thomas ran his hand over his face. "He may have been making a plea for help in some of our last conversations and the last pieces of writing he sent us."

She took his hand in hers and squeezed it. "I'm sure that your friendship went as far as it could towards convincing him that life was worth living—but it can only do so much to keep away demons like that. They're beyond anyone's control."

"But there ought to be ways of controlling them," he said fiercely.

She nodded. "I agree. But if there's anything that being bipolar has taught me, not only is evil often beyond our control, but before we can even begin to do anything about it we need to recognize it as evil."

He shook his head. "Yevgeny was never the kind to see a psychiatrist or get on meds. Not even his girlfriend could convince him to treat his depression."

"I think one of the hardest things about suicide is that it leaves the survivors constantly wondering what might have been. But wondering what might have been is only useful as a historian or scientist—not as someone living day to day in the world." She then added, "I do know what it is to feel extreme guilt in the midst of your grief, though. I've been beating myself up over the way I relied so heavily on Mom financially and the way my attachment to her was emotional in a selfish way. Meanwhile, I squandered the time she bought me."

"Your mom probably didn't see your relationship that way herself," said Thomas. "She was probably glad to do anything for you that she could. And I wouldn't say that you, of all people, have squandered time. Your many accomplishments compel admiration."

Thomas turned her lips to his and gave her a soft, slow, tender kiss. As she closed her eyes and returned the pressure, she realized she was in love with him. The certainty of her feeling, far from hitting her like a thunderbolt, emerged quietly from the shadowy recesses of her mind as if it had been waiting to announce its presence for some time. With this revelation, she felt at once exhilarated and unnerved. For her love felt forbidden in several ways. Thomas had a dark side that filled her with questions, doubts, and even fear. He didn't like music, which was not just her main passion, but her life. Further, they had only been together a short time—by Bess's own definition of love, what she felt could only be lust, not love. Even if she could overcome all these misgivings, she couldn't tell him now—not after they had both shared such deeply personal accounts of their grief.

At 11, she left on her bike to meet Albert on campus. After their practice session and her walk with Hershey, she checked email and saw that someone named Guy was interested in buying the bureau and wanted to come see it Saturday. The rest of Thursday flew by, with Melanie's and Linda's lessons followed by the boot camp. Over dinner, Nan edited some of the text of her home page. She texted Margot an hour before going to bed, and got to sleep around 10:30.

Friday morning, Nan practiced and composed for three hours, and then did further research into the SFSU bachelor's in music program in anticipation of her interview at noon. Including the director of the music school, there were four music faculty at the Zoom interview. They asked her all the questions she'd prepared for—about her teaching style and approach, challenges she'd faced in the classroom, her long-term career goals—and then a few questions about scheduling and her preferences for various courses. They needed someone to teach a music history course, a composition course, and possibly a course in the "listener's art." They asked after Nan's current compositional work, and she described exactly where she was in her writing of the quintet, saying it would be done by December 31. They asked her to send them a draft of it as soon as she'd finished.

"We liked the two unpublished compositions you uploaded in your application," said Zahr, an assistant professor in ethnomusicology.

"And your article on music pedagogy had several good insights about the value of self-reflection in both teaching and learning," added the director.

"I personally enjoyed your recording of the Brahms piano quartet," said Patrick, a lecturer in music theory.

Nan's heart rate picked up some speed. Could it be she might actually have a chance at this? They invited her to ask them some questions, which she did, and then they thanked her for the interview, wishing her a good series of holidays. "Look to hear from us again early in the New Year," said the director.

Nan felt especially fleet of foot as she ran up the fire trail just after the interview. She was full of hope for what the new year might bring, as well as for what it definitely would bring. That evening, after she'd finished playing piano at the retirement home, Nan rode her e-bike over to Thomas's bar, put the bike over against the wall, and took a seat at the far end of the bar counter.

It wasn't long before Thomas came in view, and the look on his face when he caught sight of her was worth all of her nervousness about surprising him like this: he did the quickest of double takes before a warm smile suffused his face. As he came towards her, two good-looking Black men in their late thirties waylaid him. The one with a shaved head and earring said something obviously flirtatious to Thomas, and the other one riffed on it. Thomas smiled and replied with what the first man clearly thought was a delightful response—he laughed and placed his hand on Thomas's arm invitingly. Thomas conversed easily with them both for a few minutes before advancing towards Nan.

"To what do I owe the pleasure?" he said, kissing her on the cheek.

"I just wanted to see you in another of your elements," she said. "You seem to be quite busy tonight."

"It's been an interesting evening for me. A large group of Germans came in about two hours ago. Not only do they drink generous quantities of beer, but they want to talk in German, which I'm only too happy to do." Thomas did indeed look especially pleased, and Nan found her own radiance increased by his. "How did the interview go?"

"I think it went really well, actually," she hazarded. "But of course, they probably have a slew of good candidates. I'm going to send them a draft of my composition as soon as it's done. One of the courses they'd want me to teach is a music composition class."

"Do you want to work some more on your website tomorrow after our ride?"

"I've got my students' recital until probably about 6, but I'm free after that. Won't you need to be here all evening though?"

"I should be. Why don't you bring your laptop here and we can take a table. I can get you set up with the drawing and painting software for your logo and we'll set up the subfolder you'll use for your site's images. My guess is that getting your logo and pictures looking right may take you several hours. But it's a good goal to have for the end of tomorrow."

"And then we can go back to your place?"

He smiled. "And then we can go back to my place."

He poured her a glass of red wine and, since he was pretty busy, she took out her novel to read. She read only about two paragraphs, however, before finding she was more in the mood to people-watch. A couple in their late twenties sat a few seats away from her, the woman with generous red hair that hung to her waist and the man with a sharply pointed blond goatee. Nan overheard the woman say in an accusatory voice, "You didn't take the kitchen sponge and use it in the bathroom, did you?" She couldn't catch the man's response, but he sounded vaguely defensive.

A young man in his early twenties called out to another young man who'd just entered, "Did you get the wings?"

"Dude, they were out," said the other. "I had to get nachos."

"Aw, those are wack."

Luiz caught Nan's eye, and she couldn't help recalling the fanciful scenarios Len and Carl had devised about his upcoming trip to Brazil. He came over and said, "The boss says we should take good care of you. Let me know if you need anything."

"Thank you. Luiz, who is that over there?" She nodded towards a stocky man in his fifties who was talking earnestly to Thomas and gesturing behind the bar counter. Thomas was following the man's gestures avidly and nodding at some of the things he said.

"That's the new chef—Luca," said Luiz. "He comes in every so often now to give some ideas about the new place."

"He looks as if he takes his job very seriously," Nan remarked.

"Elian says he's the best." Luiz drew off some beers into glasses and placed them on a tray. "We'll find out in a couple weeks." He winked at Nan, and started off with the tray of drinks.

A sandy-haired, red-faced man in his late fifties came and sat next to Nan, signaling to Elian. He ordered a double rye neat, and Nan noticed that he kept looking anxiously over towards the corner of the room. She followed his gaze and saw two tall, attractive blonde women in their mid-thirties conferring over glasses of white wine at a table. He had halfway finished his drink when one of the women appeared at his side and planted a kiss on his cheek.

"What are you doing over here all on your lonesome?" she chided. "Come sit with us."

The man hesitated for a brief second before standing up and following her over to where her companion sat.

Nan realized that, as Sue had hinted two weeks ago, Thomas had a goldmine in here for a novelist or a journalist—or, really, anyone who was interested in people. She wondered what the stories were behind many of the snippets of conversation she heard that evening. She pulled out her journal and jotted down a few reflections about a question she'd been thinking about for some time now: why did her greatest bursts of creative energy come just after sun had set? Most people she knew had their most active spurts just before or after dawn. She seemed to be in the minority.

An hour later, she had written this draft of a poem:

Every dusk is a miracle
The side sliver of our mind's eye
Did and did not behold the change
Stark but steady
From flame to indigo
When dreams seeped through
And an adventure was written
In the clouds that happened to streak
And striate in the pale glimmer
The spectral grade from midnight blue
To ocean azure haze reaching deep
Where we cannot ever see
Never giving the slightest hint
Of the darkness to come
The total eclipse like velvet
Over a bird's cage to lull us
Into the fast inner-woven spell
Where words not images
Approach us out of nowhere

Nan looked up to find Luiz in front of her. "Would you like another glass?" he asked.

She shook her head. "Thanks, but I think I'm going to head out pretty soon."

A few minutes later, Nan felt Thomas's unmistakable presence beside her. She looked up, smiling. "How are the Germans doing?"

"One of them gave me this." Thomas handed her a novel by Goethe called *Die Wahlverwandtschaften*. "It means *Elective Affinities*. This guy Bernd says it's pretty riveting. I guess I'll be reading some of this every day for a while. I had heard about this book when I was in Austria and always meant to tackle it."

"What's it about?"

"Apparently, Goethe's premise is that all romantic relationships are determined by chemical reactions. The question is, does that leave lovers with free will, or not?"

She smiled. "That sounds right up your alley."

"What's that you've been writing?"

She showed him her poem. "It's just a first draft."

He read through the poem carefully, and then commented, "Your metaphor of velvet over a bird's cage is really nice. The poem sounds hopeful—as if it's mostly about the 'dreams' and 'adventure.'"

She nodded. "Dusk is the time of day I always feel most energized."

"Then you should be the one working in a bar."

"Or giving an evening concert."

"Too bad you need to get to sleep early. I won't be out of here for another two hours at least."

She stood up and put her things into her bag. "I'll see you tomorrow at 10 for our ride?"

Thomas came closer and put his arms around her waist. He kissed her briefly on the lips, and then released her and stepped back.

"Oh, get a room!" said a nearby voice that Nan recognized as that of Dan Fisher.

Thomas and Nan smiled at each other, and she headed out with her bike.

♪ ♪ ♪ ♪

They had a brisk ride out Skyline Boulevard and back the next morning. The dense clouds were preparing to unloose torrents of rain. At 2, Nan met with Guy from Craigslist, who seemed to know even before he took a look at the bureau that it was exactly what he wanted. He'd come with his brother, and the two of them hoisted it, without its drawers, into his brother's pickup truck. They then used blankets to wrap up the bureau and the drawers individually. He gave Nan $300 in cash, which she promptly took over to a Chase ATM machine and deposited into her checking account. The machine took the cash, processed the deposit for a good two minutes, and then displayed a message that read, "Sorry, some or all of your deposit may not have gone through. Please call this number." Then it gave a 1-800 number. The machine printed out a receipt that repeated the same message and information. Nan pocketed it for future reference. At this moment, she had no time to follow up, since she had to prepare for her students' recital.

Nan's students shone brightly that day. They all gave their best effort and, whenever they made a mistake, they either calmly went back and played that part of the piece over again, or recovered quickly and went on. Nan could tell Neal and Shari were particularly nervous, but they overcame their nerves admirably and faced the challenge of

242

sharing their music with others. Nan was especially proud of Joanna, who gave her best rendition yet of the first and third "Dolly Suite" pieces as Nan duetted with her.

Gerald and Jess's mom had made some colorful holiday cookies, in addition to mini-sandwiches and other finger food appropriate to the five o'clock hour. Nan presented a small present to each of her students, just as Lilian had used to do—something themed at once to the student's interests and to music. Knowing, for instance, that Neal loved clocks, Nan gave him a little piano with a clock inside of it that she'd found at a store in Elmwood.

At the urging of Jess's mom, Nan put some sandwiches and other appetizers into a plastic container for Thomas. Guessing he would crave most those things that were rich in protein, she chose a selection of sandwiches with meat and lox in them. She also packed some prosciutto-wrapped persimmons, parmesan cheese straws, honey-walnut goat cheese on toasts, deviled eggs, and various small quiches. Around 5:30 people began to filter out, and Nan said her goodbyes to parents and students, congratulating the latter again on their dedication and hard work. She stopped at home to pick up her laptop, and then rode her e-bike to the bar.

Thomas's face lit up when he saw what Nan had brought for him. "I'm ravenous," he said, opening up the transparent container and popping a goat cheese toast in his mouth.

"Jess's mom said she'd made way too much," said Nan. "She's an amazing cook, isn't she?"

Thomas vouched for this by taking a large bite of a lox-and-cucumber sandwich. Nan found herself getting turned on by Thomas's undisguised hunger and her ability to satisfy it.

Between mouthfuls he said, "Let's set you up over here, and I'll show you what you'll be doing all evening."

Nan recalled herself to the task at hand. Once Thomas had showed her the steps she'd be taking to make the logo and move both it and her pictures into her subfolder, she soon became engrossed in the website. Thomas brought her a glass of white Burgundy, which she sipped as she used the drawing and painting software to design her logo. She enjoyed having the Brazilian music as a backdrop to this visual project. At first, Nan thought she didn't have enough pictures of her or her group practicing and performing. But after watching a brief YouTube video on how to capture still shots from videos, she began to pull from a few of the videos that Simon and others had made over the last year and a half.

Thomas came over a few times to check on her progress. At one point he found her deeply frustrated at the size and appearance of her logo on the actual webpage. They spent a few minutes resizing it and changing a few of the layers and effects, and then he left her to experiment further with the features of the painting software. By the end of the evening, Nan had become more conversant with inserting pictures in the midst of text and with constantly checking in a web browser to see the effects of her edits. She had presented the logo more or less as she wanted it to appear, and she had a number of photos in her subfolder, two of which were now on her main page.

"It's a good start," said Thomas, when he saw the page in Firefox. "Now, it's almost 10. Let's get you to bed."

She rode her e-bike behind him to his house, and, when they got in the door, she found herself wanting to just collapse on his couch and go to sleep right there. Thomas seemed to read her fatigue in her body language and eyes. He said, "Why don't you brush your teeth and go to bed. I'll join you in a little bit—if you're still awake."

She kissed him and did exactly as he suggested. When she next awoke, it was the middle of the night and he was sound asleep beside her. She knew she shouldn't activate her mind, but she couldn't resist taking a few moments to admire him as he slept. As she lay propped on her forearm gazing down at him, he stirred and seemed to sense her wakened presence, for he reached out and touched her hair. She lowered her face and he kissed her, drawing her in then with his hand. His eyes were closed this whole time, which somehow made his actions all the more intimate and tender. He positioned her over him and she began to kiss down his chest to his pelvic area, skating her palms along the subtle ridges of his midriff. She found herself relishing this moment in which she could do whatever she wanted with Thomas. She moved her hand along his length, kissed and licked it, and then took it in her mouth, savoring the taste and feel of him. He emitted a deep grunt that aroused Nan even more, and as she guided him into her, they began to move in sync. Images of his hand on the steering wheel, of him hungrily tossing a mini-sandwich into his mouth, and of him whipping his bike around a tight corner played in rapid succession in Nan's mind, and she broke first. Thomas followed soon after. She fell back asleep crushed against him, her head nestled in its favorite spot between his shoulder and chest.

"Can I take you out to brunch?" he asked as they lay together the next morning listening to the steady hammering of raindrops on the skylight. "You mentioned you don't have any lessons until 3."

"I'd forgotten!" She experienced the same sudden elation she'd often felt when realizing it was a federal holiday. She only had to teach Harley, Shinji, and Elizabeth later on today—all the rest had given their recital and were done for the year. "But first I need to call Bess and make sure everything's okay with her. We haven't spoken since last Saturday at the trio recital."

"That works for me. I've got to call Ken and Annie as well."

Nan was happy to hear relaxation in Bess's voice when she called her. Bess seemed glad to have the chance to open up to Nan about her conversation with Oliver the previous week.

"Oliver definitely knows that Bertie is his—I guess just from looking at him so often and putting two and two together. But I told him he can't just come barging in with his 'spiritual wonderings' and 'matters of conscience' and expect to be part of Bertie's life. I know he's lonely and bored at this point, being retired with no place to be during the day. But in the end, *we* can't be expected to fill the void. I suggested he become more active in the church, and he said he'd already volunteered to help out in various capacities." Bess sighed. "I'm hoping he finds a woman to keep him occupied."

The same thought had crossed Nan's mind before this, but she was glad it was Bess who was saying it and not she. "Where's his own daughter now?"

"Angela's a junior at UCLA in environmental marine biology. He keeps saying he wants her to meet me when she's next up here in the Bay Area—since I work for the Institute."

"How often does she come up?"

"Oh, I don't know—it doesn't really matter. Not enough. I gather her mom moved to Austin, Texas some time ago, and she usually goes to visit her mom on breaks."

"Why doesn't he go down to L.A. and visit Angela?"

"I suggested that," said Bess. "I guess he's shy of her after all these years of silence where she was clearly in her mom's camp. But since he's retired, she's made overtures towards getting to know him more— she's texted and called him a few times."

"That's promising." Then Nan added reflectively, "I certainly can understand his loneliness."

"His loneliness seems to go beyond the condition of being all alone and beyond the pandemic. He told me that once he got out of this last tour, he and his former Coast Guard friends just wanted to put it all behind them. They couldn't really stand to get together for coffees or food because what they'd shared was just related to survival. I think that's really sad."

Bess and Nan were silent for a moment. Then Nan said, "I remember how, a few years ago when I was worried I hadn't gotten married yet, Mom used to say, in the end we're all basically alone— even when we're with people in relationships, deeply in love, or surrounded by family and friends. We're all still dealing with our own pain in ways that no one else can understand, we're all still basically selfish and thinking most of the time about ourselves, and we're all still going to be alone at death."

"Mom was pretty wise. And maybe Oliver is too," Bess conceded generously.

"From what you tell me, Oliver has already seen and done a lot more than most people in a lifetime," said Nan. "He must be wise."

Thomas took Nan to a relatively new restaurant in his neighborhood appropriately called The Fledgling. It was pouring rain, and when the host asked them if they'd rather have a seat outdoors on the patio (which was thoroughly enclosed in transparent plastic) or indoors, Thomas turned to Nan. "Preference?"

Nan peered in at all the people cozily ensconced at their tables and said, "Inside."

When they had settled in at a table Nan said, "You must feel as if every time you go to a restaurant these days you're doing research for your own."

Thomas chuckled. "I'll admit it's never far from my mind."

"I'm going to just come right out and confess that brunch decisions are never very difficult for me." She picked up the menu. "If the place has fried chicken and waffle, I get that. If it has some version of eggs Benedict, I get that."

"What if, as in this case, it has both?" Thomas raised his eyebrows.

"Oh, that's going to be a challenge." Nan examined the menu. "What are you getting?"

"I'm thinking the shakshuka looks good, since it comes with olive toasts."

She took another moment to weigh the options. "All right, since these eggs Benedict have crab cakes on top of the English muffins, that clinches it."

Thomas asked after Bess, and Nan told him her sister was coming to accept Oliver more as a member of the community and not putting up so much resistance against being friendly with him. "Oliver kind of fascinates me," she said, "because the Episcopal church is so central to his life. He accepts the Christian faith as a core truth for living. I've always felt that if there is a religious truth, it must be something no one agrees on. If two or more people are in agreement about something relating to belief, it must not be true. That's why the prospect of a congregation of people getting together to worship the same idea of the same God seems really strange to me."

"How do we know they have the same idea of the same God though?" said Thomas.

"The same idea of the same God is the whole basis for a unifying tradition like the doxology," said Nan. "I gather *doxa* in classical Greek originally meant mere opinion or conjecture—as opposed to more certain truth. Then, later it came to mean popular repute. But by the time of the New Testament, this popular repute was almost solely good, and *doxa* came to mean honor and glory. So the doxology became a hymn in praise of the glory of God. In this way, an entire core ritual of the Christian church stems from a conjecture about the honor that's due to God. How can centuries of nations of people all hinge their belief on the slim framework of an individual opinion?"

"I think you're talking about two different things at once," he said. "On the one hand, you're tracing an etymological change of a word whose original sense was opinion and whose later sense was glory. On the other hand, you're finding fault with the idea of a commonly accepted religious truth that's based on mere opinion."

"I guess so." She took a sip of her coffee. "One opinion quickly develops into a whole convention. Convention and the fact that other people are doing something have always seemed to me like the very worst reasons for doing something yourself. Yet these are the bases on which people form their ideas about the creator of all things and the meaning of existence. Had I lived in the sixteenth century, I'm sure I would've been burned at the stake as soon as I could speak."

"What makes you mistrust the community to guide you in your beliefs?" he asked.

"I don't know. I guess I have a bias towards the individual's own lived experience," she replied. "If a person's convictions are formed through long, hard toil, they are probably true. If they come from a preacher's pulpit or from hearsay, they must be false."

"What about beliefs and values passed down by our grandparents and parents over the generations?"

"I tend to translate those into my own version of the truth according to my individual lived experience."

"And that brings me back to my original question. How do we know that parishioners aren't doing exactly that when they attend church on Sunday?"

She hesitated. "I don't really think that most people are capable of living in the state of distrust of conventional wisdom that I do. The world probably wouldn't function if it were made up mostly of people like me. I guess that makes me inconsistent."

He placed some shakshuka on his olive toast and handed it to her. "I think what you're really saying is you haven't made the leap from reason to faith. To a purely rational outsider, faith must always look strange. To a believer, the idea of the glory of God could never be reduced to mere 'conventional wisdom.' Maybe in matters of reason, the scientific method and other kinds of logic are required to form a plausible truth. But in matters of faith, common opinion guides most people—at least, initially. But I think you'll find that in order for them to persist in their faith, they have to make it mean something individual to them. So their ideas about God and the truth really vary quite considerably."

"That may be," she conceded. "I'd like to believe that. Do you still practice Catholicism?"

Thomas shook his head and smiled. "I guess I don't need to ask if you still practice Quakerism, for I see that you do."

"I attended meetings when I lived on the East Coast. I haven't yet gone to meeting here in the Bay Area. Partly because of the pandemic."

"Yeah, the pandemic hasn't been good for religion," he observed wryly. "And religion certainly hasn't been good for the pandemic."

Chapter Sixteen

The rain was pelting down that entire afternoon, and Thomas said it was just as easy for him to drop Nan off on his way to the bar and for her to leave her e-bike at his place.

"I'll give you the key to the front door and the code for the garage. That way, if you ever want to leave anything in the garage, or if you want to come over here before I get off work, you can."

"Wow." Nan looked down at the key. "I guess it's official then."

"What, our relationship?" He looked amused. "You really can't believe we're in one, can you?"

"It's just that it's never been this easy for me. I've always found there were tons of obstacles to be overcome and too many compromises to be made. It always ended by me deciding I wasn't willing to make all those compromises for the other person. But in this case, I wouldn't at all mind compromising—I'd welcome the chance."

She didn't add why she would welcome the chance, though the reason hung in the air for a moment: to show her strength of feeling for Thomas. She began to pack up her things.

When he dropped her off, she said, "I'll have a lot more free time now during the week, so if there's anything I can help with in the design decisions, let me know."

"Thanks. I may take you up on that. Actually, if there's a way we can do our ride next Sunday instead of Saturday, that would be great. You won't have any students on Sunday, right?"

"No. Oh, that's right, you'll have to be present during the work on Saturday. And I guess this Wednesday it's supposed to rain, so we won't get in our ride then. At any rate, Sunday works."

After her lessons ended that evening, Nan did an indoor TRX workout in her kitchen and then, after dinner, she composed for a few hours. On Monday, as torrential rains came down, she got in several hours of piano practice, mostly playing through the Franck sonata, but also playing through various Debussy pieces that often stimulated her creative juices for composing. Before she knew it, she had put aside the piano books and was working again on the second movement of her composition, this time with a firmer conviction of its overall arc. While composing, Nan wondered, not for the first time, if her discussions with Thomas weren't influencing her writing. She recalled how their exchange yesterday at the restaurant had slightly adjusted her attitude towards religion. What had her nagging perfectionism been but a one-sided monologue that pointed always towards a sealed-off, finished product? By contrast, the give-and-take of her experimental dialogues with herself as she now wrote accommodated the more flexible, feasible idea of an ever-changing work in progress. She remembered Thomas suggesting this very idea at Sue and Manuel's party, and she smiled to think that it might now be coming to fruition in her composition.

In the evening, members of the Meetup were disposed to talk about different compositional strategies, so Nan shelved the task she'd prepared for people to tackle during their hour of writing. At the end, everyone shared what they'd worked on in the hour and where they were in their process of finishing or polishing their work. Nan invited them to discuss the holiday schedule. It turned out that most members were keen to see the series through to the end, and to be there especially for the final meeting to report on their work and get feedback.

On Tuesday, the rain showers had abated enough in the late morning that Nan was able to walk over to Thomas's to get her e-bike. He told her to bring her workout clothes so they could do a workout together in his office. He showed her how to maximize her wattage output on the rowing machine, and then together they rotated among using his adjustable dumbbell set, rowing, and doing bodyweight exercises. She saw immediately how he could have so much power on the bike. He rapidly alternated upright, standing movements with lower-down, on-the-floor movements in a way calculated to make the heart rate soar. He did intervals of explosive bodyweight exercises, taking only short breaks in between. He did cardio cuts, in which he toggled between a short, intense bout of resistance work and a short period of moderate cardiovascular work. He warmed down with ten minutes of grueling abdominal exercises. By the time they had finished, in a little over an hour, they both glistened with sweat.

After they'd taken a shower, they worked on Nan's website for two hours. Thomas showed her how to tweak fonts and background colors, and how to make her text and images into clickable links.

"I really owe you," she said, when 3:30 rolled around and it was time for Thomas to get ready to leave for the bar. "How about I cook you dinner tomorrow after you've finished with the construction?"

"Sure. I'll come over to your place once everyone's gone."

Nan rode off on her e-bike and taught Allen his last lesson for the year. Then she rode over to see Bess, Bertie, and Pooh-bah. They ordered Sliver pizza in—to make up for the afternoon a few weeks ago when they'd missed it—and sat at the dining room table eating and trading stories of the rains.

"People drive way too fast on the freeways in the rain," said Bess. "I opted not to go into work yesterday because of that. There's always some three-car pile-up by 8 or 8:30, and you spend an extra 45 minutes waiting in bumper-to-bumper traffic to get clear of it."

Nan offered to take Bertie and Pooh-bah on Saturday evening after her students' recital so that Bess and Dennis could see one another, and Bess accepted her offer. The three of them walked Pooh-bah through the campus, before Nan headed back home to practice and compose.

On Wednesday Nan was able to get in a run before the next wave of rains came. For dinner with Thomas she made beef pasties in the Cornish tradition from a recipe a good friend of Gwen's. Suspecting Thomas wouldn't have time to worry about wine, she bought a bottle of red at the store down the street.

He looked keyed up when he arrived. He told Nan all about the demolition crew's work that day and the framers who'd put up most of the new frame on the kitchen.

"Tomorrow they'll finish the framing, including the new staff bathroom that's going in between the bar and the kitchen. Then Friday is the electric, plumbing, and insulation. Saturday the carpenter will put in the sheetrock. Monday the electrician and plumber will hook up the appliances and deep sink, and the contractor will tap into the gas. Tuesday the crew is following Michelle the designer's instructions for the kitchen and the new bathroom. Wednesday will be the walk-through with the city inspector. Next Thursday we'll re-open the bar while Luca meets with the new cooking staff and they start preparing bar bites to taste. By the 29th, we may be able to serve some of those same bar bites."

"Incredible," she said. "It all sounds so efficient. Are you encountering any problems with the supply chain?"

"Yeah, the price of lumber has gone up nearly 400 percent over the last few months. That's why what originally would have been a $65,000 project has turned into a $75,000 project. Also, one of the companies that Phil, the contractor, normally deals with went bankrupt during this last year of the pandemic. So he's had to go to another company and that added to the final cost."

"I hope everything else goes smoothly." Nan dished out the pieces of pasty and placed ketchup and vinegar on the table to go with them.

"Me too," he said, pulling her to him and causing her to laugh with the playfulness and surprise of his movement. "It's nice to see you."

They kissed affectionately as the pouring rain beat steadily against the windows of Nan's apartment. That night they stayed at Nan's, and the next morning Nan took delight in fixing Thomas breakfast in her own kitchen. She made up hash browns, pork sausage links, and omelets with mushrooms, bell peppers, and green onions.

"You're going to need some solid fuel for today's construction," she reasoned.

He laughed. "Well, neither Phil nor I really have to do any of the work. It's more mental than anything else. Michelle is coming in again to finish hammering out details about window trim, floor tiles, lighting, and everything else kitchen-related."

"Do you enjoy the design aspects of the project?" she asked, handing him a four-shot espresso.

"Actually, I do. I appreciate the way small decisions about the way a stove vent hangs or a countertop is designed can add up to big effects, both aesthetically and functionally."

Nan looked at her phone as a text came in, and Thomas said, "What is it?"

"Dave wants us to come to his gig tomorrow night at Schmidts Pub on Solano Avenue in Albany."

"What time?"

"It'd start around 8:00, so I'd have to bike there directly from the retirement home."

"I notice he refers to an 'us' . . . " Thomas smiled and Nan recognized he was playfully quoting her own words from a few weeks before.

"Yes, I wanted to clear up any ambiguity before we went to hear him," said Nan, "so I told him about you. He welcomes us both."

"Well, I'm afraid he'll soon detect my limitations where music is concerned."

"Yes, well, Dave may not be a chess wizard or bike racer or polyglot or well-travelled epicurean, so each to his own," she said loyally.

Thomas drew her close to him. "I guess just the way I don't fully understand your bipolarity, you can't really understand my musical shortcomings. I don't fully understand them."

"Hopefully the evening will still be fun for us both." Nan kissed him, not for the first time wondering if he felt as strongly about her as she did about him.

Thomas took off at 8:30, and Nan practiced until her therapy appointment, heading from there directly to her final meeting with Albert for the year. During their session, they made some more headway on the second movement of Franck and agreed to meet up again the first Tuesday in the new year. She walked Hershey, taught Melanie and Linda, and zipped off to the gym to give her boot camp.

Friday passed quickly for Nan, with a run up the muddy fire trail, a lesson with Charlee, and a final visit to the retirement home for the year. Fannie presented Nan with an envelope after she'd finished playing. "This is from all of us," she said.

Nan opened it and found two tickets to the Itzhak Perlman concert at Davies Symphony Hall on January 16. She broke into a smile and hugged Fannie. "Thank you," she said. "This will be the first concert I've attended since the start of the pandemic."

"Do you know who you'll take?" asked Fannie.

"Yes, I do." Nan thought immediately of Margot. "A very dear, very old friend."

♪ ♪ ♪ ♪

Half an hour later, as Nan and Thomas stepped into the sweet and funky tobacco-scented front room of the cozy Schmidts pub in Albany, Nan felt grateful for the radiant heat from the roaring fire after her frigid ride on the e-bike from the retirement home. She wished she'd thought to pick up something for them to eat on the way over, since she was, as usual, starving.

Thomas read her mind. "It looks as if we may have to sate our appetite temporarily with beer. Maybe after the music we can find some food on Solano."

Dave greeted Nan with a hug and a kiss on the cheek. She introduced him to Thomas and the two exchanged a firm handshake.

"We're about to start playing. Come on through and get a drink," said Dave.

The six musicians had set up in the back room, where dozens of people mulled around or sat talking and drinking in anticipation of the first set.

After settling Nan at a table, Thomas went to fetch them some beers. As she watched him standing at the bar talking to the bartender, Nan couldn't help marveling at his easy sociability, his cosmopolitan way of fitting in all settings with all people, and his graceful manner.

When he came back with their beers and took a seat next to her, she leaned in and said in a low voice, "Even during that fifteen minutes, I missed you."

Thomas was about to reply when the musicians started up. The group had a heavily percussive bent, partly because it featured a drum, bass, and piano, but also because the bassist led them. After they'd played through a couple of recognizable Dixieland jazz pieces, they moved on to lesser-known numbers popularized by the traditional jazz revival band located in the Bay Area in the early 1940s, the Yerba Buena Jazz Band, which was led by Lu Watters. The bassist announced all of this halfway through their set, before the group launched into some more pieces composed in the classic jazz idiom by California-based musicians. Nan kept stealing glances at Thomas as they listened, for she longed to discern what he was hearing. At one point he caught her looking at him and smiled, placing a hand on her knee beneath the table.

After the first set had finished, the musicians took a break, and Dave came over to join them. He was flushed with triumph, and looked fully confident in his abilities as he turned his beer around in his large hands.

Nan expressed her genuine amazement at the musicians' skills, both individually and collectively.

"We're recording tonight," he said modestly, "which always makes everyone perform that much better."

"You seem to have a really large repertoire," said Nan. "Or are these all standards you play consistently at your gigs?"

Dave shook his head. "No, we do try to vary up our programs, just to keep us all interested. The bassist, Sol, is a composer in his own right, and we sometimes play his pieces."

"Your drummer is incredible. He's got this steady yet driven kind of energy," said Nan with enthusiasm.

"Yeah, Sol poached him from another group in San Francisco." Dave grinned. "They were pretty pissed at us. But he lives in the East Bay, so it's a lot easier for him to play with us." He turned to Thomas. "I imagine you're pretty proud of Nan—she's an astonishing pianist, at least judging from what I heard her play the day before Thanksgiving."

Nan blushed, as Thomas clasped her hand beneath the table. He said, "To the extent that I can appreciate what she plays, I admire it very much. But I confess I'm not at all musical."

Dave seemed unfazed. "Yeah, my sister is tone deaf and she always complains that it's just so much noise I'm making."

"I like the beats I can hear—and some of your tunes were clear enough," said Thomas. "One thing about jazz, though—for me, at least—is that I don't hear the emotions in it. It all just sounds like it's going around and around without heading anywhere or saying anything definite."

Dave laughed. "You may have even more trouble with the second half of our program then, since it gets away from classic jazz and steers towards modern compositions."

One of the musicians came up and clapped a hand on Dave's shoulder. "You ready, man?"

"That's my cue," said Dave. "Enjoy the second set."

After he'd gone, Nan said to Thomas, "But you are enjoying yourself, right?"

Thomas gave a half shrug. "Sure. I guess I'm not quite as bad as Dave's sister. I can tell what they're playing is better than noise."

Nan said, "Well, it'll be over in about an hour, and we can go eat something."

She could tell Thomas wasn't listening as intently during the second half of the gig, and that he was lost in his own thoughts as he looked around the room and took in the audience. Consequently, she herself didn't enjoy the second set as much as she'd liked the first. She was rather relieved as the group closed the set after playing a dissonant number that she was fairly certain Thomas couldn't be appreciating. Part of her discouragement came from the conviction that Thomas's magnetism didn't show to such strong effect in this environment, where musicians and music reigned, and where lack of appreciation for or curiosity about music appeared as a definite weakness and, even, deficiency. She berated herself for these traitorous thoughts and tried to focus on Thomas's worldliness and cultural sophistication, which she had looked up to just before the music had begun that evening. She found images clashing in her brain—on the one hand, that of Dave confidently turning his beer around in his large saxophonist's hands and making startling and assertive harmonies emerge from his saxophone; and on the other hand, that of Thomas diffidently admitting his inability to take in the music. Nan briefly wished they hadn't come that night—that she had remained ignorant for a while longer of the extent of Thomas's musical limitations. She preferred her cluelessness at Sue's party or at the trio performance, where Thomas still seemed capable of appreciating the pith of what she'd played. With a mental thud she now began to see that she'd merely been putting off a full realization of his failure to come anywhere close to liking what she lived for.

After the applause had died down and the musicians had begun to put away their instruments, they approached Dave, and Nan gave him a hug, congratulating him on his performance.

"Thank you so much for inviting us tonight. You guys are super talented and creative," she said warmly. "I especially loved the piece you played at the end of the first half. I imagine that's a totally different instrumentation from the one the original Yerba Buena Jazz band had."

"Yeah, you're right about that. We'll be performing some more in the new year," said Dave, shaking hands with Thomas again and giving them a broad smile. "I'll keep you posted on our upcoming gigs."

As they stepped out into the cold dark night, Nan put both of her arms around Thomas and breathed in his uniquely-Thomas smell. She tried to remind herself that despite her constant affirmations to the contrary up until now, music was *not* everything.

Thomas broke into her thoughts. "Would you like to go to Zaytoon at the bottom of the street for Middle Eastern food?"

Nan's stomach rumbled in answer. "Mmmm, yes please."

They walked slowly and as one person, much as they had walked from the bar to the car on their second date, when Thomas had first shared with Nan his ideas about the long seduction. She had a renewed sense of well-being and an overwhelming urge to tell Thomas how she felt about him. But again she held back. Much as she longed to lose herself in the moment of being with him, she needed to retain enough composure to ask him some pressing questions.

They shared a plate of braised lamb shank and an appetizer of baba ganoush with an assortment of steamed vegetables and pita bread, as well as two of the cocktails for which Thomas said Zaytoon was famous. It was with great difficulty that Nan directed her senses away from the fabulous feast on the table and Thomas's engrossing presence across from her.

"I know you didn't enjoy the second half of the performance—but did you enjoy the first at least a little?" she asked bluntly.

"I enjoyed watching you enjoy it," he said generously. "I'll be honest and say that jazz gigs aren't really my cup of tea. Even during your classical concert what I enjoyed most was seeing your intensity and passion as you played. I did like hearing you improvise with the group at Sue and Manuel's that night."

"I wonder if the difference there was in the clear blues harmonies we played," she reflected.

"I'm afraid it's a stretch for me to say I really *relish* music, unless I'm doing something with my body while listening to it—like dancing or tapping out a beat or singing along with a tune—and even that ends badly."

"But that's a start," she said hopefully. "Why do you so openly broadcast to people like Dave that you're clueless in music? It may be that you'll change over time."

He looked amused, dipping a carrot in the baba ganoush. "You're holding out hope that I'll become musical in time? I wouldn't bet on it."

His light tone only spurred her to push further. "But why are you so dead-set against growing in that area? And why do you make your weakness public?"

Thomas held Nan's look steadily and she perceived in his eyes a strength of purpose that matched her conception of his profile. She ignored her instinct to revere his nobility the way she so often did, and determined to persist in her line of questioning.

He said quietly, "I could unpack a few of the elements of your last two questions. First of all, I'm intrigued by your use of the word 'weakness.'"

"You would fasten on that word," she said. "But it's true, isn't it?"

"I'll admit, I consider this a defect in my character," he said unshrinkingly. "But I want to know your purpose in bringing it up."

"I think you could be more open to mending that defect."

"So your aim is to improve me," he said coolly.

She ignored her cautious instinct and persevered. "Well, aren't you even curious about the possibilities in that area?"

"I'm not averse to being educated," he replied in even tones. "I only warn you that you're working within some narrow constraints—among other things, a forty-year-old who has several competing passions and limited time, not to mention little natural proclivity in that direction."

"As a teacher, I don't shy away from challenges like that," she said. "If you have the will, that's most of the battle won."

"Let's examine something else you just said," he said, taking a sip of his cocktail. "You seem to be troubled that I'm honest about my 'weakness' in public. Why is that?"

"Well, people like Dave or my trio cohort don't need to know that you don't understand music." She flushed.

"So you're ashamed of my defect."

"You're not?"

"Why should I be?" he said. "I'm human and therefore fallible. I'm sure there are a few things Dave doesn't understand that I do. For that matter, there may be a few things you don't understand. But I don't hold them against you. And I wouldn't be ashamed if you decided to talk about them amongst acquaintances."

"Now you make me sound unforgiving and relentless," she said, abashed.

Rather than pressing the point, he returned to a previous thread. "Why do you assume I'm 'dead-set against growing' in the area of music?"

"Only because you keep harping on your limitations."

"It seems you would be content to deny them." He met her eyes shrewdly.

"No, never that. But I would focus more on education and less on your drawbacks."

"What are you ultimately aiming for, Nan?"

"I guess I just want a sign that you could like a little what I live for," she said.

"What if I said that it might never be possible for us to share together, the way you share with your friends, the passion of music— of concerts, recordings, and compositions?"

"I would say that's a tragedy."

"Is it a tragedy that our relationship can't survive?" he put to her frankly.

"I don't know," she said with equal honesty. "I've known a few couples who were mismatched in this way. In graduate school I had a friend named Teresa who was brilliant at clarinet and for whom music meant the world. She ended up marrying a mathematician who was tone deaf and who couldn't share her love of music. They've been going strong for eight years and they have two kids. She's very much in love with him."

"How did they make it work?" he asked.

"She just came to accept the fact that she can enjoy music with her friends and colleagues, while she and her husband enjoy other things together."

He looked at her meaningfully. "What's wrong with that?"

"It just seems like such a huge compromise," she said, not daring to meet his look, but lowering her eyes to his hands.

He paused, and then said in a quiet voice, "Did you ever consider that compromise might work the opposite way from how you fear it works?"

Intrigued, Nan looked back up at Thomas's eyes, which had retained their dynamic glimmer. "How do you mean?"

"You assume that compromise weakens the person who compromises—or that it lessens his or her love. But supposing you never had to compromise when caring for someone close to you. I posit that you would end up feeling far less keen love for that person than if you have to compromise in order to be with him or her. It's the compromising that seals your affection and gives you a strong bond with the other person."

"But how can that be? It seems so counterintuitive."

"What you put effort into is what your heart cherishes," he said simply. "To draw a pedagogical analogy you might appreciate, don't you ever feel closer to the students who stumble under your care and who need extra guidance, than to those students who breeze through your courses acing all the assignments, who neither need much help from you nor ask for help?"

Once more, she was struck not just by his sharply analytical mind, but by his far-reaching imagination. "Indeed, it is very much so. But I don't know how you could deduce this from your own experience."

He ignored this as he drove home his point. "And isn't it the work you put into those foundering students that gives you the greatest reward and sense of achievement?"

"Undeniably so, yes."

"So why wouldn't you put similar effort into someone you love?"

Unable to keep from smiling, she conceded to him. "I suppose you're right. Seen from that angle, it seems so reasonable."

"Well, I'm not sure it's reasonable, so much as an odd facet of human character," he said, pushing his plate aside and folding up his napkin. "Nature abhors a waste of energy."

♪ ♪ ♪ ♪

Nan felt more than a little insecure as she rode her e-bike behind Thomas to her place later that night.

She pulled up behind him, parked her bike, and came over to the driver's-side window.

"Have I ruined anything between us by questioning you tonight?"

"I'm getting used to it—in fact, I've come to expect it," he said noncommittally.

Nan supposed he must be comparing her ambivalence tonight with all the times in recent weeks that she'd doubted they were in a relationship, and she flushed with guilt.

"Only, I feel I may have pushed too far," she said, still not daring to meet his eyes.

"It's true that tonight our sparring hit a little too close to home."

"I hate for us to part without complete understanding." She finally looked him in the eye.

"That may be a bit too much to ask for. How about a little better understanding than we had before?" To her relief she saw a spirited spark in his eyes.

"But you're not mad at me?"

"How could I be? I hope you realize by now, I'm happy to learn from all our conflicts."

"What have you learned from tonight's?"

"I've learned I'm not good enough for you," he said with a defiant glint in his eye—and a trace of something else that she couldn't identify.

"Please don't tease me with that," she said, ashamed. "I'm sorry I'm only slowly learning to be more flexible. And then my usual perversity kicked in and I had to rock the boat."

"It's not surprising you should want to test the limits of our relationship. But since you made trial of it, I might also ask you, what have you learned?"

"I've learned that however self-knowing we think ourselves, we can always learn more about ourselves," she suggested. "And that

260

weaknesses can become strengths when two people make a commitment to one another."

She was leaning both her forearms on his open window. She smelled the alcohol on his breath and it activated her dormant desire: she longed to taste his mouth, and to let him know with her lips, at least, how strongly she felt about him. She kept hoping he would take hold of her hand or place his hand on her arm, but he kept his hands in his lap.

"Those are good starting points," he observed. "But only experience can bear out the truth of either reflection."

She was as frustrated by his indubitable wisdom as by his continued aloofness. "But won't you come in for a cup of tea? Or stay over for the night?"

"Not tonight. Tomorrow the sheetrocking starts at 8 sharp. I'll see you Sunday morning for our ride."

She hesitated, and then said softly, "You may not miss me, but I'll miss you."

He seemed to read in her eyes the need for reassurance, for he leaned over and kissed her. The kiss didn't last as long as she could have wished, but she did take comfort in it.

"I'll miss you plenty," he said. "For now, goodnight."

Nan stepped back as he gave a wave and put the car in gear, making his usual one-point U-turn and driving back towards Oakland. As she watched the car disappear in the distance, she mulled over what Thomas had said earlier about compromise. She couldn't remember the last time she had wanted to compromise for anyone. For some reason, however, she had mentally and emotionally committed to him almost from their first date, thereby ensuring that she would need to compromise. Why had she done that if not to test her own ability and willingness to give a little? He had effectively suggested that in loving someone you especially loved their weaknesses. Until now, this idea would have been anathema to her. But she determined to sit with it and give it due consideration.

It was strange, on Saturday morning, not to be riding with Thomas. She did a backyard resistance workout instead, and then prepared for the piano recital at Elizabeth's home in Upper Rockridge. The recital had an even more festive vibe than the one of the preceding week, since it was now exactly a week before Christmas. Everyone brought some kind of dessert or drink with a holiday theme, and Linda's husband Ron brought a rich *bûche de Noël* from a bakery on the

Northside of Berkeley. All of Nan's students acquitted themselves well in their performances, and talk was animated afterwards.

At 6, Nan headed back home to change for her time with Bertie and Pooh-bah. She checked her email and saw a message from one of her credit cards saying, "We missed your payment this month. Let's make this right." She thought quickly. The only way the automatic payment wouldn't have gone through was if there was no money in her checking account on which to draw. A leaden weight dropped in her stomach as she opened her bank statement and saw that not only had she been charged an overdraw fee, but she had a negative balance of -$95. Then she remembered the deposit slip she'd received the previous Saturday. With the greatest difficulty, she pushed all of this to the back burner yet again as she went over to Bess's apartment. She would deal with it tonight after Bertie had gone home.

As she, Bertie, and Pooh-bah walked back to her place from Bess's that evening, on the narrow sidewalk of College Avenue in Elmwood a well-weathered homeless man approached them shouting out all manner of obscenities and threats at the volume of a professional opera singer. Pooh-bah began to growl fiercely, and Nan saw there was nowhere for them to go unless they turned back and crossed the street at the crosswalk. She turned to lead them back that way, when the homeless man, who was in his late fifties with a scraggly beard and wild eyes, began to lash out and advance towards them, making gestures as if he was going to punch Nan and kick Pooh-bah. Nan's heart rate sped up and her adrenaline kicked in. She shouted, "Get the fuck away from us!" Meanwhile, she prepared herself to hit or kick the man back if he attempted to hurt them.

The man made a feint as if to head-butt Nan, and then at the last minute turned and shuffled on, muttering loudly and turning back once or twice with a loud curse.

Bertie said, "That was a scary man."

"I'm sorry, Bertie." Nan was mortified that she had cursed in front of him—that she had cursed at all. They went back to her place, and she tried to calm down. But the thought of attacking the bank problem later loomed heavily, and she wasn't able fully to enjoy her time with Bertie and Pooh-bah. Bess came and collected Bertie around 10. She and Dennis had had dinner in North Beach and walked around a bit afterwards.

"We're off to Nevada City early on Tuesday," said Bess. "Bertie, Pooh-bah, and I are flying back to Minneapolis Friday afternoon."

"I leave then too." Nan made an effort to appear absorbed in her sister but sensed she was failing miserably. "I couldn't believe how expensive tickets were this year."

When she'd said goodbye to Bess and Bertie, Nan's barely contained burst of adrenalin returned, as she finally hunted up the deposit receipt from a week ago and called the 1-800 number. While she pressed the numbers, she was already shaking with anger and resentment that the bank had taken her cash without crediting her. She'd worked so hard to get that bureau sold and to put that money in the bank in advance of the credit cards' debits. Over the course of the first half-hour of waiting on the phone, she had to speak to about three people before she was connected with a woman in Bangladesh named Baaji.

Nan tried actively to remain calm throughout the entire slow process of question and answer that followed, but it took monumental effort on her part. Baaji kept asking things first that Nan couldn't decipher unless Nan asked her once or twice to repeat them. Then it seemed Baaji was asking the same questions over and over again at various intervals, so that Nan was pretty sure she'd answered the question "How much did you deposit?" or the question "Which bill denominations did you place in the bill holder?" about five times. It seemed Baaji was repeatedly testing Nan to see if she was telling the truth. Randomly, in the middle of the hour and a half-long process, Baaji asked her, "How are you doing today?" and it was all Nan could do to suppress her rage. But she knew that lashing out would only prolong this endless torment. After a pause she managed to respond, "Very well, thanks." But she was seething, especially since Baaji kept placing her on hold "for just two minutes," which invariably turned into five or six minutes with some sterile commercial music playing in the interim.

In these between-times, when she was on hold, Nan started muttering to herself. "What kind of torture have they devised here?" "I can't wait to get hold of the bank manager Monday to argue with him about the overdrawn fee." "I'm going to get the credit cards to refund me the late charge. They've got to refund me." "Chase can't get away with this. If they're doing this to other people, they can't function. "This is going to take all of Monday to sort out." "The internet, electric, and gas must've also tried to collect the money and failed."

At last, Baaji said that her supervisor had "okayed" a "temporary reversal" of $300 in her account, which should show up in 24 to 48 hours. "Is there anything else that I can assist you with today, Nane?" she asked.

"It's Nan. No thank you."

Nan pressed end on the call, and, without thinking, screamed out, "FUUUUUUCCCCCCKKKKKKKK!!!"

She looked at the microwave clock and saw it was 11:40. She had ruined her bedtime and wasted a huge block of time, she would need to waste even more time on Monday to get back the money that was rightfully hers, and she'd been nearly attacked by a homeless man. She had to slam something, so she hurriedly put on her shoes, opened her kitchen door, slammed it twice with great huge bangs, and stomped down the stairs. She punctuated all of this with a few long screams, as she walked down to the bottom of the driveway and out the gate.

"I'm going to play piano all night long. I might as well at this point—I'm not going to sleep anyways," she announced loudly. She decided to play the most crashing passages of Beethoven that she could think of—the "Appassionata" sonata, the third movement of the "Moonlight" sonata, the first movement of the "Pathétique," the "Waldstein." She would show the world what it was to fuck with her time. It was quite cold outside, but Nan felt so hyped-up that, jacketless as she was, she was actually hot. She walked around doing the same voice warm-ups she did before she sang opera music, and doing this in as unbridled a way as she could so that passers-by who were coming from late nights out turned to look at her before hurrying in the opposite direction.

After aimlessly walking to the top of College Avenue in the cold, Nan returned, slammed the door several times again, and took a seat at the piano. She played through all of the tempestuous Beethoven passages, and then, feeling more satisfied, but more manic than ever, she decided to work her way through Book Two of Bach's *Well-Tempered Clavier*. By then it was 3 a.m. and the fatigue that Nan had felt after her students' recital had completely worn off. She was so keyed up that she began to hear, in Bach's music, invective against the powerful of this world.

She paused and took up her notebook, jotting down some notes for a poem against banks in which she meant to address the fact that nothing of what they did was risky anymore—that they were guaranteed to pocket everyone's money with impunity—that they bowed and scraped to their wealthiest clients—that they were largely above the law. She wrote out bits and pieces of this poem until 5:30 a.m., at which point she heard some rattling outside near the trash cans. She dashed downstairs to find a homeless man riffling through their recycling bin for bottles and cans. She went up to him and said forcefully, "Don't come here this early in the morning to do that. You'll wake someone up. And this is private property." She waited until he had shuffled off full of mutterings before she ran back upstairs.

Nan realized, as she returned, quaking, into her apartment, that she was being all too inconsistent. Here she had likely kept a few people in her own apartment house awake, and she was upbraiding an innocent man for making a little noise while merely trying to survive. Further, she was filling up pages with words inveighing against the rich and powerful, and then, in her actions, turning on someone with no power at all. But she was in no mood to hold herself accountable to consistency.

At 6, Nan was exhausted and beginning to hear voices in her head— or were they outside of her head? First a crowd of voices would come at her from the left, starting with a low murmur and then swelling to a cacophony. Then they would ease off as individual voices would chime in from the right. She could make out the voice of an ex-boyfriend, Gabe—with whom she'd kept up on Facebook—who was describing to her the three most perfect crime schemes now being committed in the world today, and did she want to join in on one of them and make some money? Then the goose-like voice of an unidentified old woman squawked on and on about how Nan would never be able to afford living in her old age if she didn't start saving and investing now. Waves of babel rose and fell like this with an insistent clamor until Nan, in desperation, shook her head violently, placed her hands over her ears, and with intense effort repeated over and over, "Quiet the mind. Quiet the mind." She played some of the first movement of her quintet on the piano to drown out the voices, and gradually they died down completely.

At 7:30, Nan decided there was no point in going to bed. Thomas would be over at 10, and it would be all the more painful to wake to the alarm after only two hours of sleep. She determined to power through their ride on no sleep. She hoped that the worst of her manic bout had subsided, as she tried to regain her composure so that no obvious traces would remain of her episode.

But Thomas was too shrewd for her. When he came in to greet her as she was suiting up for their ride, he took one look at her and the apartment and seemed to read in them both the story of her night's activities. She had been in no mood for cleaning up: poetry books lay scattered all over, notebooks remained open on her desk, the piano books she'd played from were on the floor, and the layers of clothes she'd cast off in her extreme overheated state were strewn about carelessly. When Nan went manic it was as if she took extra delight in being as slovenly as she could possibly be. Even the kitchen had dirty dishes and pans all over that Nan had meant to clean up after Bertie and Pooh-bah left; but she'd called the bank instead. She suspected that her eyes still bore the same wild aspect as her apartment.

Nan could almost hear a low whistle passing through Thomas's mind as he surveyed the damage, before he said quietly, "What happened?"

She told him everything, starting from the previous Saturday to the night before going into this morning. He listened to the entire story without saying a word. Then he advanced towards her and placed his arm around her back as he said, "You must be very tired now. I can head out alone while you get some sleep . . ."

"No!" Nan didn't mean to let so much alarm infuse her voice. She added, more calmly, "I think this is going to help to go out on a ride. There'll be plenty of time later to sleep."

"Won't you be too weak to climb though?" He looked skeptical.

"If the past is anything to go by, I'll actually be more powerful today," she said.

She hastily put on the rest of her cycling gear and they headed up into the hills. The cold was raw and damp, and Nan, who hadn't eaten much for breakfast, felt keen hunger; but as they rode, she devoured power bars, GU shots, and electrolyte chews. They rolled to Lake Chabot via Marciel Road, off of Redwood, and then returned back up Pinehurst. A heavy mist still clung to the roads, for there was hardly any wind, and Nan cried out joyfully when she spotted a lean and shaggy coyote observing them with glowing eyes from a grove of scrub oaks and bay laurels. She wondered if he had breakfasted yet, or if he was already hunting down his dinner. The mountains were expelling their rills into the creeks of the canyon, and puffs of aromatic sagebrush blew gently into their path at unexpected intervals. It was the sort of morning where one didn't miss the absent sun, but basked in the quiet pallor of hidden things emerging out of the grey. This monochrome scene appeared especially distinct and memorable, and Nan felt an enormous surge of love for Thomas. He was the only person she wanted to be with this morning, and he was the only person who was more than worthy of her.

They flew down Claremont, stopping only once they'd gotten back to Nan's place.

"You took all the hairpin turns pretty fearlessly today," said Thomas. "And you're right. You were extra fast, as if you'd either gained sudden power or lost some weight—or both. I guess I'm lucky you don't have manic episodes more often."

As they stood at the bottom of Nan's driveway, she felt full of restlessness still.

"You're not mad at me, are you?" she said distractedly.

"Why would I be?"

"For not better managing my bipolarity." Nan looked down at the ground.

"All things considered, it sounds as if you have managed most of it." Thomas placed his hand under her chin and tilted her face up to him. "But now, I think you need to finish off the job by getting some sleep."

Nan nodded. "Will I see you later?"

"Call me when you wake up, and we can have dinner at my place."

Nan thought of something. "I've just got to go to Peet's coffee to get a bag of beans since they'll be closed later when I get up."

"I'll come with you. Take your bike upstairs first, and we can walk over there."

When they reached Peet's, which was just a block down from Nan's apartment, there was a line inside, so they paused for a moment by the wrought-iron tables outside. As was always the case on a Sunday afternoon, some laid-back young students had spread out all their gear across the tables and were working on their laptops and using their cell phones for a number of tasks. Suddenly, out of the corner of her eye, Nan saw the flash of two bodies on either side of her snatching and grabbing all the electronics in sight on the tables. Two young adults— barely seventeen, Nan would have guessed—darted with their spoils in their arms towards the open door of a beat-up grey car that was waiting ten feet from the curb. Nan's shock turned to sudden determination. Without thinking, she lunged after them in the street, and as they began to close the car door, she reached for the door with the aim of wresting some of the electronics from the arms of one of the thieves.

As she tried to prevent the car door from closing, two strong forearms closed around her waist and pulled her back, during which delay the young man slammed the door and the car sped off.

"That's never a good idea," said Thomas as he continued to guide her towards the sidewalk. "For one thing, you don't know if they have a gun. And for another thing, desperation lends strength to even the weakest fighters."

Nan shook with rage. "But how can they get away with this? Has anyone called the police?"

Two witnesses to the snatch-and-grab raised their phones and said at the same time, "They're on their way."

Nan turned to Thomas disbelievingly. "Did you see how swiftly and confidently they carried it all out? They must've been circling the block several times before this to scope out their target."

He nodded, but his eyes remained fixed protectively on Nan and he kept his arm around her as the ensuing action unfolded. The students were upset, but it hadn't really sunk in yet how much they'd lost. Only when the police arrived and began to question them and all the witnesses did the students begin to realize how slim the chances were that they were going to retrieve any of their electronics. The young woman turned pale and said shakily, "*All* my work was on that laptop. And it wasn't backed up."

The officers shook their heads with blank expressions. "We've got some cars tracking down the license plate you took a picture of, but the perpetrators will probably change the plate or even dump the car. All they want is to offload the electronics and get from the sale a mere fraction of what the items are worth. They're probably going to make a total of $500 from this."

Needless to say, this was cold comfort for the poor students to whom Nan's heart went out. Through all this, a kindly Peet's barista came out and calmly distributed among everyone present—including the officers—a complimentary free drink card. "On the house," she said evenly. "Just our way of saying sorry." Clearly this was far from her first snatch-and-grab.

Meanwhile, more people had congregated on the corner and the witnesses retold their story for the fifth time to freshly eager ears.

Eventually, when it seemed there was nothing further Nan or Thomas could do, Thomas said, "Let me walk you back home."

Nan didn't question anything he did or said. It had slowly dawned on her over the last half-hour that he had probably just saved her life. She didn't know what she might've done had she been allowed to pursue the thieves as she'd intended to do—in her rage she probably would've tried to fight with them. But she now suspected that the outcome could have been very bad indeed.

"I should thank you," she said.

"You certainly are fearless when you're manic," he observed. "It's a little frightening."

"You're so steady through all this." She looked at him with newfound awe. "How do you manage to keep your cool so easily?"

"It's a ruse. Underneath, I was afraid for you."

"All the more impressive that you were able to remain level-headed. I could have been seriously hurt. You saved me from myself."

"I don't want to lose you just yet."

With one hand Thomas pushed his bike, and with the other he guided Nan back to her apartment.

He said, "I'm going to come back here after taking my bike home. I'll stay with you while you take your nap. Will you be up for the next half-hour or so?"

"Yes," she said. "But it's not necessary. I'll just be sleeping."

He shook his head decisively. "I want to make sure you're okay."

True to his word, he returned and parked his car in under half an hour. He hadn't taken the time to shower at his place, but had brought clean clothes in which to change out of his bike clothes. He used her tiny shower, about which she apologized the whole time, but he didn't comment on its deficiencies. As she climbed into bed, he sat on top of the covers next to her with his Goethe novel in hand. He translated an interesting passage to her aloud, and she listened to the reassuring murmur of his gentle voice, dropping at last into a deep sleep.

She slept from 2:30 until 6:30, and when she awoke, it was as if the mania had mostly passed. She felt, if not restored, then at least refreshed.

Thomas was upstairs using her laptop when she emerged.

"How are you feeling?" he asked.

"Much better." She kissed his cheek and wrapped her arms around his waist. She didn't realize until then how glad she was that he had insisted on remaining by her side. It was unprecedented to have someone care for her like this in her manic state. "Thank you for staying. I'm sorry I took up your entire free day."

He gave a soothing laugh. "I wouldn't have passed it any other way than how we spent it."

She rubbed her eyes. "I think the monster passed while I was sleeping. I'm fit to be seen again."

He took her hand and held it between his two palms. "For your sake, I'm happy to hear it. But don't ever worry that I'll be frightened off by the monster."

"I'm afraid it brings a bit more discord than even you and I are used to."

"Then I really have no fear of it."

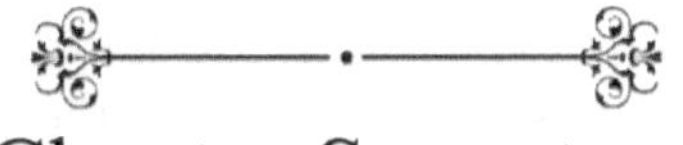

Chapter Seventeen

Thomas grilled up mahi mahi tuna that evening and served it with rice and grilled asparagus. They sat, as usual, side by side at his bar counter.

"I figured you'd be starving after your sleepless night, our tough ride, and then your long nap," he said, pouring them glasses of sparkling white wine. Then he added, "It's just a little bit humbling to ride with you when you're manic."

She laughed. "'Humbled' is not the way I would normally describe you, so that's really saying something."

"I've got my fair share of self-doubt," he said soberly. His expression, meanwhile, was surprisingly guarded and uncertain.

She became serious to match his demeanor. "Is all of this modesty stemming from my manic episode?"

He seemed to consider his words carefully before speaking. "I suppose there's always the danger, in creating a work of art, that one can become the work of art too fully—that there's nothing left of oneself. Much as I love where we are and where we're headed in the long seduction, I wouldn't want it to become so all-consuming that we leave no room for taking a perspective on it."

"You mean we risk becoming our roles too narrowly—with no room for anything beyond them?"

"Exactly." He placed his hand over Nan's and traced her veins with his thumb, looking down at their hands rather than directly at Nan as he spoke. "I'm fully aware that if there are stars in relationships, you ought by all rights to be the star of ours. You're incredibly talented on many fronts, well educated, self-knowing, and sexier than you know."

Blushing, she interrupted. "Well, what about you? You've got one of the sharpest minds I've met, an intimidating will, a noble spirit, abundant self-awareness, and supreme hotness to top it all off."

He smiled slightly, still averting his gaze as he continued to stroke her hand. "I know that most of the time our roles come all too naturally. I feel simultaneously the pleasure of the conquest and your pleasure at being overcome. Two of my greatest delights these days are to remind

271

you that I can overpower you and to remind you that you belong completely to me and me alone. But I want you also to know that through all this I, too, experience doubt. These same pleasures sometimes seem precarious, and I savor them from a position of relative insecurity.”

“Didn’t you say that doubt was inherent to the long seduction—that it helps to create the necessary suspense, on the one hand, and the glue for the relationship, on the other?”

He nodded. “For a moment, though, I’m inviting us to step outside of our roles long enough to examine who we are as we embody those roles—if that isn’t too much of a contradiction. There’s no question that by keeping you guessing, I can keep the tension high and the danger live. I know you’re receptive to any and all cues I give you on that front. But there are unknowns for me in all of this as well—a question of whether or not I’m worthy, hesitation at being always in command, and a sense of vulnerability as our relationship deepens.”

She said quietly, “And here I thought I was the only one to have constant questions and doubts.”

He shook his head. “The other night, for instance, during and after the music performance, I felt unworthy of you. I doubted my ability to keep you.”

“That was my fault,” she said with remorse. “I was too harsh a critic and too unremitting a judge. I’m only gradually learning to temper those tendencies.”

“Your words at dinner merely confirmed my own thoughts. I’ll admit that for much of the evening I was even jealous of Dave.”

She was floored, even as she felt secretly flattered. “I never would have guessed.”

“I’m afraid it’s too deeply ingrained in me not to show my jealousy. Jealousy smacks too much of weakness.”

“But does that mean you just keep it to yourself?” she said with disbelief. “Doesn’t that make you resent other people?”

“No,” he said. “I hold it against myself. Even now, I’m risking a lot in telling you.”

“But you did emphasize how important jealousy is to the long seduction . . . ”

“I know,” he said with a grimace. “I guess I forgot how inconvenient—how painful—jealousy can be.”

"I've thought a lot about it since Friday, and I realize you and I have so many things we share together—cycling, exercise, food, drink, art, literature, travel—we don't need to share music as well."

"No, but I won't deny that it felt like a moment of defeat for me losing the spotlight to Dave in an area in which I can never excel—an area to which you devote your life."

"Was that why didn't you want to stay over that night?"

"Yes. And I wanted you to be able to think through everything we talked about over dinner—to be sure you did want to continue in our relationship without the haze of sex to cloud your judgement."

She squeezed his hand gently. "I'm glad you've spoken plainly, and I hope you feel free enough to do so at any point. But how is it you can manage to tell me all of this with such graciousness and candor that your image in my mind only grows more perfect?"

His brow remained furrowed. "As long as you remember it's only an image. And no, you're not alone in having doubts."

"Since we're speaking so frankly, I wonder . . . " She hesitated, and Thomas met and held her gaze, as if encouraging her to continue. "Well, I sometimes worry that if it weren't for the fact that I love everything we do towards the long seduction—including our power roles when we make love—I might not hold your interest."

He bowed his head for a moment, and then, when he lifted it, she saw that the softness his eyes had held the entire day since her bipolar episode had begun had deepened. "If being with you so far has taught me anything about myself, it's that the attributes you have that drive you to prefer the long seduction are what I love—your curiosity, your energy, your playfulness, your fearlessness, your intellectuality. Take all of these away and yes, perhaps—" here Thomas smiled— "I would be less captivated. But when push comes to shove, I believe I am the one who should worry more on that score."

"Why do you bring up your doubts now, just after I've gone manic?"

"Lately we seem to be seeing each other at our most vulnerable. I wanted to seize on this opportunity to make an honest confession to you about the extent of my own exposure."

Nan thought for a moment. Her resurgence of love for him now, when he had laid bare his weaknesses, surely demonstrated more than just the fact that doubt solidified the bond between two lovers. Surely it also suggested that her own love could embrace his imperfections. That would be a striking novelty. She turned it over slowly in her mind and considered it from several angles. Accepting imperfection was beginning to feel less and less like settling. Thomas had just unquestioningly guarded her all day in her weakened state—only a day after he had steadily weathered her irresolution over their compatibility. The contrast in their approaches to the whole character of a person hinted at a previously unknown capacity of love to fashion defects into a harmonious picture that pleased for its very defects. Thomas's confession to jealousy and self-doubt suggested that he was ahead of her, not only in loving her but in loving her for her imperfections. Should she lay her cards on the table, risking everything, or hold back? She decided to wait.

♪ ♪ ♪ ♪

"I really need a steady job," said Nan early the next morning, after they had made love and were tangled up together in Thomas's sheets. "I can't keep cutting it so close with finances."

"You mentioned possibly doing a year-long stint teaching music in a public high school under some permit?" he suggested.

"Yeah, the Substitute Teacher Incentive Program, or STIP." Nan ran her fingers over Thomas's firm bicep. "I was putting that off because you only receive one of those once in your life per state, so it's best to start it at the beginning of a school year, when you can get a whole year's use out of it."

"Would you enjoy teaching high school music?"

"I would be directing bands, orchestras, and choirs, and I'd probably be moving around schools within one district like Berkeley Unified," she said. "It would certainly be challenging, because everyone needs to wear masks at all times in schools, and that's detrimental to playing any mouth-based instrument, including singing. It's also really hard to understand what students are saying and to make yourself understood by them without a microphone."

"And you wouldn't have that problem at a job like the SFSU one?"

"Half their classes are online, which simplifies things. Also, they're not playing mouth-based instruments in the courses I'd teach. I could probably use a mic to lecture in person."

"Do you have any other college-level jobs in the offing?" Thomas initiated a playful dialogue of toes with his and Nan's feet that caused her to giggle.

274

"I apply to two or three every day," she said. "Community colleges, four-year colleges, universities—I'd be glad of any teaching gig just to get my foot back in the door."

"You seem to have feet on the mind," he teased, tracing his own shapely foot up her calf and thrilling her with the sight and feel of the gesture.

She groaned, and he moved over her.

"You were saying, about applications?" he prompted, holding down her hands as he tantalized her further by tickling her groin with his own length.

"What?" she said, struggling in vain to recall the drift of their conversation as he continued to muddle her mind with his seductive movements. "It's . . . Ohhh . . . "

"Maybe you'll remember later," he said, pretending to lift off of her entirely.

She instinctively tried to raise her hands to stop him from going, but he kept her firmly pinned down. She ached to have him inside her, and her hips arced off the bed with longing for what they had nearly had.

He smiled down at her, watching her writhe for a moment, before he slowly eased back down over her and trailed his sex lingeringly downward over her belly. "Are you looking for this?"

She twisted in eager anticipation of his entrance and let out a great sigh of agony. Instead of entering her directly, he bent down and pressed his lips to hers. Glad to have even this to fasten onto, Nan reached for his tongue with her own—but after each of its brief incursions, it always managed to retreat just beyond her grasp. Her mind filled up with a sense of Thomas's complete power over her at this moment, and she sighed again, this time in surrender to his superior force.

"You're just as beautiful when you struggle as when you capitulate," he mused as he watched her facial features resolve themselves into yielding. When he finally entered her, she took each purposeful stroke as a declaration of how he possessed her entirely.

It wasn't long before they both erupted in cries and groans, and Thomas loosened his hold on Nan's hands. She clasped his hands more tightly and found herself whispering, "I love you."

Thomas was still coming, and she couldn't be certain whether he'd heard her—nor did she venture to repeat it for fear that he might have heard her. She did and didn't want him to know what she'd accidentally let slip.

Over their brief breakfast, he said, "I'm getting my booster shot tomorrow at eleven. I figured I'd better get it over with before things start heating up at The Chaser."

"Won't you have side effects that get in the way of overseeing the work?" she asked.

"I doubt it. I never really had side effects from either of the first two shots."

They both headed out at the same time, Nan on her e-bike and Thomas in his car.

That day was filled with a flurry of mundane tasks. Nan argued her way out of all the fees incurred by the Chase bank incident. She did two large loads of laundry in preparation for her trip on Friday and thoroughly cleaned her apartment. She did a brief shop at Berkeley Bowl, where she bought a few quarts of soups to match the cold weather. She ran up at the track, where she caught up with an 86-year-old regular named Tim, who jogged slowly but persistently around the track eight times every other day. She also managed to compose for two hours after dinner before preparing for an early bedtime. Instead of texting Margot tonight, she called her.

"How are you, love?" she asked when Margot picked up.

"Okay," said Margot. "I had a mediocre date on Saturday with that guy Noel."

"Oh, the actor?"

"Yeah. He's nice enough. But I'm just not sure I felt the spark."

"When do you leave with Christina for Tahoe?"

"We're renting the car Thursday evening and driving up early Friday morning."

"I guess we'll have to get in our hill walk in the City after the New Year," said Nan. "I have a little something for you too—I'll give it to you as a Twelfth Night present."

Once again, Nan slept a solid twelve hours, waking after 10 the next morning. She felt much lighter and more clear-headed than she had in a long while. She lay in bed ruminating over what she'd inadvertently told Thomas yesterday morning. Had he even heard her? If he had, was he avoiding the subject? What were his own feelings? She realized now how awkward her position was. Maybe he hadn't taken her seriously because she'd gone manic the day before she'd said those three momentous words—he may have doubted she meant them. This was a fine situation to get herself into just a few days before she was taking off for a ten-day absence from Thomas!

She tried to put all of this aside in her mind as she went through Facebook, Nextdoor, and other sites renewing her listing as a piano teacher and replying to messages people had sent her. She'd managed to secure a walk with Hershey in the midday, and this time Calvin Liu had requested she make it two hours. "We'll be going out of town on Wednesday," he'd texted, "and we want him to be nice and tired out for the trip." She led Hershey up the steep Claremont Canyon fire trail, where the heavy traffic of hikers after the recent rains had wrought stark new ruts in the dirt. As they descended via the gentler UC Berkeley fire trail, Nan breathed in deeply to catch the fragrant traces of eucalyptus, bay laurel, and pines. From the top of the steep connector she surveyed the span of Berkeley and Oakland shimmering like a white-topped labyrinth whose grooved arteries pointed towards the bridge, the Bay, the City, and, beyond that, the ocean. What did it matter, she thought, if Thomas did or didn't respond to her declaration right away? She knew it was the truth, and the truth needed to be spoken. Her heightened sensitivity to everything in the hills today testified to the strength of her own feelings.

Just as Nan was preparing for bed that night, around 9:45, Thomas called.

"How did the booster go?" she asked.

"I thought it went okay . . . until now." His voice sounded shockingly weak. "It seems to have hit me pretty hard after all."

"Shall I come over?"

"I just feel incredibly fatigued. And I seem to have a fever and chills. When I lay down after coming home, I couldn't get warm. But then I woke up and had sweat through my shirt completely. The sheets are soaking."

"I'll be over in just a few minutes," she promised.

She took two of the soups from Berkeley Bowl out of the fridge and packed them into a bag, together with a liter of soda water and a box of herbal tea bags. She hurriedly threw some night things into another bag and loaded up her e-bike. She arrived at Thomas's ten minutes later and, after putting the bike in the garage, let herself in the front door with the key.

She cautiously opened the bedroom door a crack and saw that Thomas had one of the bedside lamps on. He stirred, and Nan approached the bed hesitantly.

"Thanks for coming," he said. "You didn't have to."

"You sounded pretty bad. Are you hungry or thirsty? I've brought some soups, tea, and soda water."

"I think I just need to sleep for awhile. But listen," he added in a tone that suggested he was making an extreme effort to form his words coherently.

"Yes?" Nan sat down gently on the side of the bed, brushing his moist hair off his forehead with her fingers, and awaiting any instructions.

"Could you possibly go do the walk-through with the inspector tomorrow at The Chaser? I can't get Michelle on such short notice—she's got a few jobs going at once now."

"Of course. What will it involve?"

She noted a few beads of sweat on Thomas's brow, and she longed to go fetch a washcloth, run it under cold water, and place it on his forehead.

He spoke as if from a great distance. "He'll go over the plumbing, electrical, gas, the fire system—and make sure all the fixtures and appliances are up to code. You just need to be there to listen and watch as he fills in all the details on the papers. He may have special notes or questions—you don't have to have answers, but just hear him out."

"Okay. I can do that."

"The keys to the place are on the nail by the fridge in the kitchen. The walk-through starts at nine sharp."

"I'll be there." She went into the bathroom, ran a washcloth under cold water, and returned to Thomas. But by the time she got back, he'd fallen into such a fast sleep that she couldn't bear to wake him.

She stood for a moment contemplating his beauty, and then she stepped out of the room to put the soups away in the fridge. She brushed her teeth and joined him in bed, turning off the light.

By early morning the sheets were soaked through with his sweat. Although Nan wanted to change them, she knew it would take five minutes that Thomas was unable to spare. At 8:30, she took a last look at him as he slept soundly next to her. She got out of bed, dressed, took the keys to The Chaser, and left on her e-bike.

The inspector, who introduced himself as Brian, seemed only momentarily surprised to find Nan leading the walk-through instead of Thomas. Phil, the contractor, showed up ten minutes late, when Nan and Brian had already walked through the new staff bathroom and part of the kitchen. Brian posed various questions to Phil about the spacing between the edge of one of the counters and the walk-in cooler, a few outlets at the back of the kitchen, and a shortened support beam in the roof where Phil had had to work around an overhanging tree outside. Nan listened carefully to everything they said and committed it to memory, in case it should be necessary later. In the end, Brian gave The Chaser a "pass," and said that he'd be sending the official pass to Thomas in the mail.

"Good luck with your enterprise," said Brian to Nan.

Directly after Brian had left, Luca arrived to take a look at the kitchen. He passed through, shaking his head and grunting with disapproval at several things before he turned on Phil. "I know Michelle and Thomas did not tell you to put the cooler door like this. How can we get in and out to the dumpster without bumping into each other coming out of the fridge? What about these shelves—these are for pans but we're gonna need the pans over *here*, not there. This is where the dishes should go. What is this prep counter doing here under the food warmer? Whose idea was that?"

Phil had a simple answer for everything: "This is what they told us to do."

Luca wasn't satisfied. He continued to critique everything from the placement of the garbage receptacles to the color of the paint on the walls. "What kind of white is that—it looks like a hospital in here."

Nan kept silent, trying not to let Luca's criticisms get under her skin. Surely when Michelle arrived later, all of these things would get sorted out?

Luca left in high dudgeon, leaving them with the portentous words "You'll see just what a disaster it will be when I try to cook in here with the new staff tomorrow."

Nan looked over at Phil, who merely shrugged and said, "Cooks. There's no pleasing them. I have to take off for another job, but have Thomas call me if Michelle or he wants to change anything." And just like that, Phil disappeared.

It was only 10:30 and Michelle wasn't due in until 1, so Nan rode back to Thomas's place to check on him. All was quiet when she arrived. She warmed up some soup, made some tea, and moistened a new washcloth with cold water. When she came into the bedroom, Thomas was half awake.

"That smells good," he said.

Nan sat on the side of the bed and felt his forehead the way her mom used to do when she was sick. He was still burning up, and she placed the cool washcloth over him. She noticed that the sheets were still quite damp. "Shall I change the sheets for you? It won't take more than five minutes."

Thomas didn't say anything, but slowly got out of bed and went shivering over to a chair, where Nan wrapped him in the duvet. She quickly took off the old sheets and replaced them with fresh ones she'd found in the linen closet. She then gently urged Thomas back to bed and tucked him in, replacing the duvet over the top of the sheets. She took the old sheets out to the laundry area, and when she got back, Thomas was sitting up slightly.

Nan placed the tray of minestrone soup and tea on his lap. Thomas had only a few bites of the soup and a sip of the tea before he seemed to want to return to the business of sleeping. She placed the tray on the nightstand, kissed him on his hot cheek, and left him.

She realized only later that he hadn't even asked about the inspection.

She went back to The Chaser at 1 and met with Michelle, who was a radiant Black woman in her early forties with curly red-highlighted hair and a warm smile. Her very presence reassured Nan that the kitchen scenario was not as dire as she'd thought when Luca was there earlier. Michelle listened to everything Nan reported and nodded. "These are all things that can easily be fixed if Luca wants them that way. Give him one or two nights of cooking in the kitchen and we can iron it all out, no problem. Otherwise, it seems everything went off without a hitch. That's excellent."

Michelle made a few notes on her iPad as she walked through, and said she'd send off a more detailed email to Thomas. "Tell him I hope he feels better. My own booster was pretty bad too."

When Michelle had left, Nan pretended that she was a cook at The Chaser. She went through the kitchen and admired the brand-new cookware and utensils, the pristine cooler, the shiny new fryers, the spotless countertops, the as-yet unused sinks. She was in awe of the fact that the entire construction crew had done all this in only six days. She especially loved the way the bar opened directly onto the kitchen so that customers could see the cooks in action.

Nan rode back again to Thomas's and set her e-bike charging in the garage. Finding that Thomas was still asleep, she went into his office and stretched on the floor mat for a while. Then she pulled out her notepad and worked on the cello and viola parts of her second movement. She sat on the couch writing as the afternoon sun passed down into the southwest corner of the living room. A little after 5, she decided she would light a fire to combat the extreme cold. She opened up the flue, placed some kindling and paper on the grate, lit a match, and got the base of the fire started. She found some logs in the garage and brought them in. Before long, a cheerful fire blazed in the fireplace, and she stared, mesmerized, into the flames.

"I haven't used that in a year," said Thomas's voice behind her. As always, she thrilled at the sound, however weakened it was by his sickness.

"How are you feeling?" She turned to him and placed a hand on his forehead. He was less warm this time.

"A little better. This is the booster shot from hell."

"Did you get something to eat and drink yet?"

"I'm going to bring your tray in here and have it by the fire," he said. "Since you've made it so pleasant in here."

Nan cleared off the couch so he could lie on it, and she tucked him in with the blanket on the chair. "Are you warm enough?"

He nodded. She rewarmed his soup and made him some fresh tea, and then came and kneeled by him on the floor, both of them gazing into the fire. She was surprised at the simple pleasure she took in being needed by Thomas and being able to fulfill his needs. She felt a more insistent pull towards him than ever before.

"How did the inspection go?" he asked.

"You passed. Luca came in and made a stink about the kitchen, but then Michelle came in and assured me everything would be okay."

He nodded again. "That sounds about right."

"Are you going to be able to open the place again tomorrow evening?"

"I think so. I just need one more night of sleep—and sweating it out."

"Imagine what actually getting Covid would be like, if this is what getting the shot is like," she reflected.

"I've been thinking about that," he said, staring into the flames. "Once again, I feel lucky to be basically strong."

"Now's the time I should challenge you to a game of chess," she said with a mischievous side glance. "I might actually stand a chance while you're sick."

"Don't count on it." A glitter returned to his previously lusterless eye. "Chess is another game entirely.

Chapter Eighteen

By Thursday midday, Thomas was able to get out of bed and have some of the vegetable barley soup Nan had brought. He was well enough to stand at his office desk and read through the news and check emails. Nan rode her e-bike behind him as he drove to the bar. She had canceled her therapy appointment for today.

"Tell me what I can do," she said, once they were inside.

Thomas thought out loud. "Elian's the only bartender tonight and tomorrow night, so I'll be helping him out with drinks. Luca and the new cooks will be messing around in the kitchen from 5 onwards. Chris the barback comes in at 4 and starts setting up. We won't put the 'Grand Opening' sign up till Luca gives the word about the bar bites. I guess if you could be on hand to help deal with any problems in the kitchen, that would be great. Maybe just listen to whatever Luca suggests and jot down notes—and reassure him we'll take care of whatever he wants as soon as we can."

Nan was glad she didn't have to give her boot camp that night. So many people were gone for the holidays that the gym had canceled it.

She soon found that the work at the new restaurant provided much more exercise than boot camp ever could have. Outside, the construction crew had put up the new name, "The Chaser Restaurant and Bar," along with the cycling logo, and repainted the front of the place. This new exterior, together with Thomas's recently updated website, the announcements he'd made on social media, and word of mouth drew in dozens of new people. The night was incredibly busy. Luca arrived around 5 with a van full of food supplies, which Nan and Thomas helped unload into the kitchen. By 6 the cooks were all busy preparing the sample bar bites for the staff to try. Nan spent most of the evening in the kitchen helping Luca's cooks move various items and hearing out Luca's complaints, which she duly noted down to pass along to Thomas.

Around 8:30, Nan helped the cooks set out the various sample bar bites at a counter on the side of the kitchen, and she laid out plates and silverware for everyone. Elian and Chris came in by turns to try a few.

Thomas eventually came in as well, and Nan noticed that Luca stopped what he was doing to observe Thomas's reactions to each taste. Thomas turned to her and said, "Nan, you'd better try these as well." He smiled at Luca. "Her opinion matters just as much as mine."

The spread was truly amazing. Luca and his new staff had prepared an olive medley; thick-cut garlic fries with aioli; a selection of pickled vegetables; salt-cod croquettes with a green sauce; fried polenta with chanterelle mushrooms, gruyère, and caramelized onions; Baja fish tacos; and bacon-wrapped sriracha buffalo wings with a blue cheese dip. All of them had distinct but well-harmonized flavors, and the variety of the tastes was no less extraordinary.

"We can add more meat bites in later," Luca declared. "For now, we're gonna focus on fish and vegetables."

Nan and Luca both noticed that Thomas especially loved the fried polenta and the buffalo wings. "So when do we start serving these, Luca?" he asked, snagging another cod croquette.

"If you can get the supplies in by the 29th, we can serve these the three nights leading up to the New Year," said Luca.

"So we could have the Grand Opening on the 29th?"

"Sure." Having established that Thomas was happy, Luca passed on to other things. "We need to train the wait staff tomorrow, though. And you wanna have menus of some kind ready."

Thomas turned to Nan. "Let's get those finished up now."

For the rest of the night Thomas went back and forth between serving people drinks and checking on the menu Nan was modifying on his laptop at the bar. Thomas provided her with all the prices and the main ingredients of each appetizer, and she phrased the descriptions as temptingly as she could. She had seen the way menus listed lower-priced items at the top and the pricier items towards the bottom, so she arranged the items similarly. She tweaked a few of the fonts and colors in the same way she had done with her own website. They printed out the menu in various sizes, and finally settled upon a 6 x 10-inch light-grey card stock with magenta and black lettering. The cycling logo was displayed up in the right-hand corner across from the restaurant's name.

"We'll worry about printing out the drinks menu at another point," said Thomas. "This is a good start."

"What about scannable QR codes for the menus?" she asked.

"Later," he said. "You've got to crawl before you walk."

She stayed at the bar until 10:30. Since the kitchen crew were all gone and there didn't at present seem to be as much needing doing, Thomas encouraged her to go home and get some sleep.

"Do you want to take one last ride up over Grizzly Peak tomorrow?" he said. "I can drive you to the airport."

"Thank you. That sounds great." She felt his forehead one last time, and nodded to find that it was a normal temperature. Then she leaned in and kissed him. "I hope you can go home soon and get some sleep yourself."

When she got home, she realized she had to pack now, since she wouldn't have time after their bike ride tomorrow. She put on some music from the archives of the jazz radio station and dashed around the apartment gathering clothes, watercolor materials, toiletries, music books, her laptop, and gifts for her family. It was with difficulty that she was able to stuff everything into the suitcase and her backpack, but at last she was able to zip up both.

Christmas Eve morning was bitterly cold, and, when Thomas came up to Nan's apartment, she wrapped him in a long, secure hug. "I'm afraid when we stop hugging I'll go back to shivering," she breathed into his neck.

"Wear another layer on your head," Thomas advised. "That's where most of your heat escapes from."

As they climbed the hill, Thomas told her about the last two hours at the bar the previous night, which had apparently been busier than the bar had ever been. "Tonight should be even crazier," he said. "Luca's bringing over the ingredients to start experimenting with one or two of the main courses. The new wait staff will also be coming to get trained by Luca. At least we'll have Shelley as a second barback to help Chris out."

When they got to the highest part of Grizzly Peak, Nan looked to the left and noticed how clear Mt. Tamalpais showed across the Bay. She slowed up.

"Can we get a picture of us here in front of the view?"

They stopped at the huge logs and rocks where people routinely took photos. A couple of friendly twenty-something guys, who looked and sounded high already, agreed to take their picture. They took several shots, up close and further back.

Nan showed Thomas the pictures on her phone, and he smiled. "You look far better than I do. I look sick in all of these."

Nan objected. "You look great in all of them!"

At that moment a large yellow-tailed hawk alighted confidently ten feet away from them, surveying his surroundings as if considering where to find his lunch. Nan and Thomas exchanged admiring looks, and Nan took a picture of the lofty bird.

They rode to the bottom of San Pablo Dam Road, turned around, and rode back through Moraga to climb Pinehurst. They nodded at fellow cyclists who were similarly getting in their Christmas Eve rides before the holiday demands of family and friends precluded riding.

Thomas came back to pick Nan up at 1 sharp, and he lifted her heavy suitcase into his trunk with relative ease. Nan put the jazz station on, and they slowly made their way onto the Bay Bridge, which was jam-packed with cars headed into the City, many of which, like them, were going to the airport.

As they sat in traffic with the jazzy holiday tunes playing over the car speakers, Thomas looked over at Nan a few times. He seemed to want to say something, but then stopped himself each time before actually saying it.

Finally, Nan laughed. "What is it?"

"You said something the other day, I think, after we had sex."

"Oh." Nan's stomach fluttered as she realized that this was the moment of truth. He *had* heard what she'd said after all.

"I just wondered if that wasn't the mania talking?"

She tried to gauge the tone of his question. Was he asking her this so that he could have an escape route, or did he think that she had inadvertently cheapened the three-word phrase, or did he suspect the truth and want them both to lay their cards on the table?

She decided to interpret his question in the last sense. "It wasn't the mania talking. I was telling the truth."

Thomas stared out at the lines of cars stretching into the City. At last he said simply, "I love you too."

"You do?"

He shook his head, smiling. "Always doubting."

"But I'm crazy, and insolvent, and high-strung, and selfish . . ."

"Only in ways that I love. I wouldn't take you any other way."

"How long have you known?"

"Since that night you took the bath and I suggested tying you down and you got so excited."

Nan feigned offense. "So if I'd said no, you might not have fallen in love with me?"

"I said that was the night I *knew*. It had been a while in coming before that."

"But why did you know that night?"

"The moment just reminded me of all the reasons I love you—which I did enumerate the night after your bipolar episode, as you'll remember."

"I knew when we were having breakfast in your kitchen and you were so sympathetic about my mom's death—and when you told me about Yevgeny. Something inside me clicked into place and I realized you had all the qualities essential to a hero—generosity, courage, and compassion."

"I'm sure my unstable representation of that ideal will give cause for much doubt—in both you and me," he observed.

"Well, you can remain a comic hero," she said playfully, taking hold of his free hand. She added, with pride, "One whose restaurant-bar is going to be the hottest spot in Uptown." Then she said wistfully, "Now I really don't want to go back home. Maybe I can send a doppelgänger in my place."

"It's only ten days."

"You must not love me as I love you, if you can say that," she said reproachfully.

"I'll show you how I love you—by giving your bike a good washing while you're gone. It needs it badly."

At the drop-off curb they kissed for as long as they could before the curbside security officer began to hustle Thomas along.

"I'll text you when I get in," she said.

As Nan sat at the gate waiting to board the plane, her thoughts were full of Thomas. She imagined him working the bar tonight until 1 and then getting on a flight to Baltimore at 6 a.m. tomorrow, and she prayed his immune system was strong enough to travel after so much work and exercise and so little sleep. She pulled out the pictures of them together on Grizzly Peak from that morning and sent them all to Thomas with a message that said, "Who would ever guess you'd just gotten your booster?"

She knew this period of transition from being with him to suddenly not being with him was inevitably the hardest of all, but she couldn't help longing to take an Uber back to Oakland and stay with him for the last fifteen hours before he left. In order to keep the tears from coming, she put on her iPod and listened to Mendelssohn's piano quartets, which never failed to cheer her up. She texted Margot and asked how Tahoe was.

"There's so much snow here!!!" Margot texted back. "It's unbelievably beautiful."

"You and I will both have a white Christmas," Nan wrote. "How is the house?"

"Gorgeous. We keep the fire going constantly, and Christina's brother James is so cute and sweet. He's starting up the grill in a few minutes and making rib-eyes."

"Keep me posted on the James front . . ." Nan texted.

Margot sent back a series of heart-eye emojis and images of a steak and a barbecue.

When the plane was taxiing to the runway, a text came in from Thomas. "Forgot to say, last night I hung up your portrait of me behind the bar to inaugurate the new place. Got many enthusiastic comments about it. Dan F. said you made me look way better than in real life."

Nan burst out laughing, and sent a series of laughing faces before turning off her phone for the flight.

As was always the case, coming into the Twin Cities airport on a snowy evening in December was dramatic and glorious. The white blanket of snow that covered everything except for dark ribbons of road and the fine brown tracings of trees reflected the lights of the plane and shone with a purple twilit glow. Nan always found she adjusted quickly to the Midwestern cold, and tonight was no exception: as she stood outside on the curb waiting for her dad, she found the below-freezing temperatures invigorating. Frank picked Nan up in the blue Subaru he and Gwen had owned for twenty years now. They hugged long and tightly, and Nan hastened to put her bags in the trunk before Frank could do so.

"I've made roast chicken tonight." Frank pulled out of the airport and onto the freeway. "Bess, Bertie, and Pooh-bah will be coming in an hour on Delta, but Bess wants to take an Uber. I had to go out now anyway, to get some things we'll need for tomorrow's dinner. Everything will be shut tomorrow, of course."

Nan asked about Frank's classes at the University and he told her of the two lecture courses he'd be teaching in the upcoming spring semester. "One of my PhD students will be coming tomorrow for Christmas dinner with his wife—Theo and Elise. Their families are in Vancouver, Canada and they just didn't feel like traveling that far this holiday season."

"You look really good, Dad. Are you still using the elliptical machine regularly downstairs in the basement?"

"Every day," said Frank. "And now, as you suggested, after finishing the elliptical I do a few dumbbell exercises. They've made all the difference for my balance and posture."

"I can show you lots more than just the exercises we did together this summer," she offered.

When they came up the basement stairs to the kitchen, Frank stood back to look at her. "Uh-oh. I can tell you're in love."

She blushed and smiled. "You, Bess, and Margot are too good at reading me."

"Tell me all about him while we put these potatoes through the mandolin. I decided to do scalloped potatoes instead of mashed tonight. But first, go ahead and take your things upstairs."

Since Gwen had died, Frank had been channeling her spirit by cooking for hours in the large kitchen where Gwen herself had devoted so much time to preparing food for nearly forty years. While cooking didn't come as naturally to Frank as it had to Gwen—he was far more painstaking and meticulous in his methods—it gave him great joy to think that he was making Gwen present by engaging in one of her greatest passions in the room where together they had talked, laughed, and shared the pleasures of good food and drink for their entire marriage. Frank never considered that time wasted that was spent planning out and executing a delicious meal.

As Nan unpacked upstairs, she thought of how difficult Frank's own journey of grief had been since her mom had died. For the first year and a half, he had been full of bitter despair and anger. Loneliness had settled in soon after, and while Nan was glad he had sympathetic colleagues and students to engage him, she also hoped that one day he might be able to date again. He was a youthful 72-year-old—healthy, physically active, intellectually sharp, good-natured, and fired by the same kind of curiosity for all things that stimulated Nan herself. Like Nan, since losing Gwen, Frank had struggled with regaining a sense of humor about the things that before her death had readily assumed a comic aspect. He was only now beginning to take an interest in entertaining again.

Dinners at home were often quite late, in the European tradition, and Frank made dry martinis for him and Nan to sip as they chatted and cooked. Teddy and Naomi came over around 8:30, and Bess, Bertie, and Pooh-bah arrived at 9. Naomi was about four inches shorter than

Nan, with straight dark brown hair and a slight frame. Teddy was about six feet and thin, with his reddish-brown hair cropped short. The large kitchen was abuzz with conversations, and Frank, Nan, and Bess stepped nimbly around everyone as they completed the dinner preparations. Pooh-bah, who was especially tired from flying, heaved a great sigh and went directly over to his bed by the wood-burning stove in the corner. Bertie was punchy and disposed to run around the various rooms of the downstairs at high speed.

Bess shrugged. "We'll let him work off his energy. He'll sleep all the later tomorrow."

"How was Nevada City with Dennis?" asked Nan.

"Really nice. Please tell Doug thank you again, Dad—we left a note and a present for his daughter. On Wednesday afternoon we went to the Nevada City Winery and tasted more than a half dozen of their wines. Dennis and I both ended up buying assorted cases. We all walked out the old Downieville Road on Thursday. The shops in town are so eclectic. Most of what we got for all of you comes from what we found there."

Both Teddy and Naomi were rather quiet people, so Nan was glad to have been placed at table between both of them. As the family passed around the root vegetables and other dishes, Nan asked Naomi, "Isn't your sister Rebecca coming into town for a few days?"

"Yes," said Naomi. "She's driving in tomorrow from Chicago."

"She's bringing her African grey parrot Augusta." Teddy served himself some chicken, passing it along to Bess on his left. "Named after Aunt Augusta in *The Importance of Being Earnest*."

Nan laughed. "Is she imperious and unyielding?"

Naomi took a sip of soda water; neither she nor Teddy drank much alcohol. "She has a voice that sounds just like Lady Bracknell's in one of the T.V. performances of the play. My sister has had her for seven years now."

Teddy added, "She's really intelligent and has a great sense of humor."

"Your sister or the parrot?" joked Nan.

"Both," said Teddy and Naomi at once, laughing.

"Nan, where's Thomas spending the holiday?" asked Frank.

Nan lit up from within. "He's flying out early tomorrow morning to be with his older sister Annie and her family in Baltimore."

"You mentioned he loves keeping up his languages." Frank served Bertie some chicken. "I'll give you a book he may enjoy that the

University of Minnesota Press just published on how people learn and retain languages."

"He's recently started reading a novel by Goethe," said Nan. "He translates some of the more interesting parts aloud to me."

Frank mused, "As far as I can tell, a lot of those popular language apps like Duolingo and Rosetta Stone discourage users from translating the target language back into their native language. This is unfortunate, I think. Translation not only makes language learners realize the differences between the foreign and the native languages, but sharpens their skills at switching back and forth between the two languages. It also activates more parts of the brain."

"Teddy, are you still using Duolingo to learn Russian?" asked Bess.

"Yeah. I may actually be able to talk with some of Naomi's Russian cousins at the wedding." Teddy helped himself to another serving of potatoes.

"Dennis has also used it to bone up on his Spanish." Bess took a spoonful of root vegetables. "I guess I'm the only one who hasn't given it a whirl. Maybe in the New Year I'll start learning Portuguese."

"I like Portuguese," Bertie announced very distinctly in the silence that followed, causing everyone to laugh.

After dinner, they all gathered around the fire in the living room. Bess had brought a Jamaican Christmas cake that featured marzipan, almond-flavored icing, and fruitcake with rum- and port-soaked fruits in it. Frank served port with the cake. Bertie took one look at the cake and said, "Yuck."

Frank chuckled. "Bertie, I have some special Christmas cookies for you."

Nan had put on some old, well-loved Christmas CDs of the Boston Camerata singing early American songs, the Dutch mezzo soprano Ely Ameling singing Christmas music from Europe, and the San Francisco-based group Chanticleer singing medieval and Renaissance holiday songs. Thanks to Teddy and Naomi, the Christmas tree was decked with all the traditional decorations that Gwen had put on every Christmas tree from time immemorial, as well as multiple strings of colored lights that winked back merrily through reflection in the dark windows. At the top was the same glass angel that had always perched atop the tree, blowing his bugle in a perpetually still and listening night.

Frank raised his glass in a toast. "To young love," he said, "without which old love would have no meaning."

Bess and Nan smiled at each other, clinking glasses with Teddy and Naomi.

♪ ♪ ♪ ♪

Christmas morning dawned so clear and bright that the sun glancing off the snow dazzled the eyes. Nan, Bertie, and Pooh-bah tramped down to the Minnehaha Creek, which was less than a mile away. Kids of all ages sped down well-worn snow slopes on sleds, toboggans, and flat pieces of cardboard, their shrieks filling the air and drowning out the occasional sound of dogs barking at the excitement. At first, Pooh-bah was enchanted by the opportunities that the snow afforded for burrowing. After awhile, however, he began to lift his paws in puzzlement and look up at Bertie and Nan as if to say, "You got me into this; now, can't you get me out of it?"

Nan picked up Pooh-bah and slung him in her jacket. "Poor Pooh-bah. We need to buy him some of those booties that Midwestern dogs have."

Bertie took endless delight in flinging large snowballs onto the thick ice that covered much of the creek. They found some spots where the black rushing current was too strong for ice to form, and Bertie's snowballs dissolved in the cold water. At one point they saw a muskrat swimming along upstream.

"Do you think he has his home somewhere here?" asked Bertie.

"Let's follow him," said Nan.

"We need to look for thick piles of grass," Bertie said knowingly.

They walked upstream until they had lost sight of the muskrat; but they came upon a family of skiers who were coming in the opposite direction.

"Merry Christmas!" the parents called out as they passed Nan, Bertie, and Pooh-bah.

"Merry Christmas!" echoed Bertie.

"Bertie, you're a true Minnesotan," Nan remarked. "You take to the cold naturally."

"I love the snow," said Bertie. "Later can we can go see the ducks?"

Mallards always flocked to a particularly warm and sheltered area of the creek where dark waters remained unfrozen even during the bitterest days of winter, and Bertie loved to watch the ducks bathe, sleep, waddle, and quack as though it were high summer.

"We will," Nan promised. "First, let's go eat something."

291

When they got back home, they found a fire lit in the kitchen stove, and a plate of ginger lemon muffins made by Bess with a note saying, "For Bertie and Nan."

Frank sat at the kitchen table absorbed in the newspaper.

"Where's Bess?" Nan started up the coffee maker.

"She found a store that was open until noon and went out to get some last-minute supplies," said Frank absently.

Nan texted Thomas. "Would you like to Facetime at 5 p.m. your time?" She knew he wouldn't get the text until a couple of hours later—he was probably flying over her right now—but since she had a Mac and he had an iPhone, she figured this might be the easiest way for them to catch up.

Bess got back just before Teddy and Naomi came over, and everyone congregated around the stockings that hung by the kitchen stove. This had always been the way in the Arbuthnot family: they opened stockings on Christmas morning and Christmas tree gifts on Christmas evening before dinner. Then Naomi and Bess helped Nan and Frank stuff the goose, put it in the oven to roast, and prepare the gravy. At 4, Nan excused herself and went to the quiet office on the other side of the house to Facetime with Thomas.

Thomas picked up after two rings, angling the phone so that Nan could see his sister Annie and his three-year-old nephew Martin waving in the background. "Leo's somewhere here," said Thomas.

"He went out to get more booze," Annie corrected him.

Nan smiled. "Do you guys have places still open there in Baltimore?"

"A few liquor stores that can't resist making a killing on Christmas," said Annie, who bore a striking resemblance to Thomas, Nan noted with pleasure.

Thomas came back in full view. "How's Minneapolis?"

Nan told him about the sledders and skiers, the runners, fat bikers, hockey players, and kiteboarders.

"That sounds like a pretty active city."

She nodded. "It's almost as if the colder it gets, the fitter the people in the Twin Cities become."

"What will you eat for dinner?"

"Dad's making a stuffed goose for the first time ever. He's bringing out the usual assortment of wines from the cellar. And I think Naomi baked a couple of pies. What about you all?"

"Annie's making a standing rib roast with carrots and mashed potatoes. I'm in charge of the salad dressing, and I gather Leo is making the cocktails."

"How lucky are we?"

They talked a bit about the bar's booming business of the night before, and then he asked how close she was to finishing her composition.

"I need to work steadily on it for the next six days. I think the end is in sight," she said, tempering her excitement with caution. "I just wonder if future compositions are going to be this hard."

"Don't think about that yet. Just get this one completed."

"Do you guys have any special holiday traditions?" she asked.

Thomas looked around, as if checking that no one was within earshot. "Leo is really into zombie movies at Christmas, for some reason. He's liable to make us watch a few later on tonight."

She laughed. "You'll probably become a zombie yourself while watching—you must be dead tired."

"I am. But it's good to be away for a few days."

"I miss you," said Nan.

"I miss you too," said Thomas smiling.

"Merry Christmas."

"Merry Christmas."

When they hung up the call, Nan thought what a letdown mediums like Facetime or Zoom always were when you really wanted to *be* with a person. They neither allowed the mind to roam free and imagine what the other person was doing, nor did they allow the body to touch the other person and be in their presence. She felt she'd rather have Thomas's shirt to breathe in than a screen projecting his image two-dimensionally. But at least there was still his soft, mellow voice . . .

"How's Thomas?" Bess's own calm, alto voice sounded from the doorway, and she came in holding a mug of tea.

"Tired, I think," said Nan. "He had his booster the other day, and it took its toll on him."

"Yeah, I know someone who got appendicitis from her booster." Bess sat on the couch. "It's a pretty serious dose of mRNA. For some reason, a lot of people are getting thrown for a loop by this shot."

"How's Dennis?"

Bess's face lit up. "He's really good. We had a lot of time on our own once Bertie had gone to sleep." She lowered her voice. "And Bertie likes him a lot. One day he said, 'Can Dennis be my dad?' It was a little awkward, because Dennis was right there making toast. I said, 'What made you think of that, Bertie?' He shrugged and said, 'A lot of people I know have dads.'"

"Where did the conversation go after that?" asked Nan, breathless.

"I said, 'There are different kinds of dads—some who are there only when you're born, some who are there your whole life, and some who come into your life later.' 'Which one do I have?' he asked. I said, 'The last kind.' He gave that some thought, and then said, 'How late will my dad come?' I said, 'When you're a little older.' He seemed okay with that."

"Whew! That's some serious conversation to have over toast." Nan exhaled at last.

"Dennis just smiled and said under his breath to me, 'Mine was the first kind, and that's probably for the best.'"

At 5 Rebecca arrived with Augusta the parrot, just as Theo and Elise came. There were hugs all around, and everyone gathered by the Christmas tree to unwrap presents. Bess gave Nan a dark blue, fleece-lined exercise shirt for cold runs and daily wear around the house. Teddy gave Bertie a puzzle that involved different configurations of solid cubes needing to be fit together into a whole cube. Nan gave Bess a stunning wool scarf that blended several shades of aquamarine that perfectly matched Bess's eyes. Naomi gave Frank a pair of draftsman bookends made out of iron calipers used for engineering and mapmaking. Frank's face glowed as he turned the bookends over in his hands. "What a cunning gift," he said.

When everyone had opened their gifts, Frank said, "Now, I must get back to the goose. I don't want to overcook it. I need to put the cranberry-filled apples in to bake as well. Nan, can you help take the stuffing out of the bird so I can carve it? Naomi, if you could finish the gravy, and Bess, could you put out the Brussels sprouts and dress the kale salad."

They all repaired to the kitchen carrying their pre-dinner drinks. Pooh-bah eagerly danced around the feet of anyone who bore meat, cheese, or cream, and Augusta the parrot avidly observed the collective stir from her perch on Rebecca's shoulder. Every so often she would mutter something like "Oh, this is gonna be big" or "Don't go there!" or "Does this make me look fat?"

Rebecca explained. "That last one comes from her previous owner of many years ago. I'm afraid because we always laugh when she says it, it's still stuck with her."

Frank made a small pile of pieces of goose and mixed these into Pooh-bah's dog food, causing Pooh-bah to leap out of his bed and attack his bowl with glee. Everyone carried the various platters and bowls into the dining room, and Frank poured the first bottle of wine, a Beaujolais Gamay, into their glasses. Nan had lit four long candles, and she dimmed the overhead light so that everyone appeared in warm candle glow. Bess helped Bertie to those dishes she knew he'd like, as well as a few dishes she hoped he'd take to.

When they'd all filled their plates, Frank said, "We're very fortunate to be where we are, with each other, only a year after things seemed so dark and unending. Let us hope the next year brings similar dramatic turns for the better across the entire world."

This brief grace made tears start at the edge of Nan's eyes. Then Augusta the parrot bellowed, in the exact voice of Lady Bracknell, "Merry Christmas, everybody!" returning the table to laughter.

"Pass me the gravy, Nan?" said Bess.

Frank speared a Brussels sprout on his fork. "Bess, I'm so happy that you and Dennis could enjoy Doug's place in Nevada City. I'm thinking of spending a few days there in early June before coming down to see you and Nan in Berkeley. Tell us more about the policy work Dennis does."

Bess explained, "He works with the research team, the scientists, me, and the rest of the stats group. Our last project was part of an ongoing series on combating 'carbon lock-in,' which is the situation our country, like most others, has gotten itself into of being deeply dependent on fossil fuels. Dennis addresses policy changes that remove some of the technical, social, and political hurdles to taking action. Even though his work applies mostly to California's climate change policy, it's cited by researchers and policy makers across the world."

"He must be incredibly busy if he also does consulting," Frank remarked, passing the stuffing over to Teddy.

"He's eased off of that for a while." Bess smiled, and Nan could see her sister loved talking about him, just as she loved talking about Thomas.

Frank asked, "What field of research are you in, Rebecca?"

"I'm a research audiologist at Northwestern," said Rebecca. "I develop and test hearing aids and then help write up our findings."

"How wonderful!" Bess poured some more wine into her glass. "We'll all be benefitting from your research in a few years' time."

"According to some of my students, I could benefit from it already," said Frank.

Nan asked, "Haven't they developed some pretty good apps that link up cell phones to hearing aids for people who want to listen to music in different environments?"

Rebecca nodded. "It used to be that people with only moderate hearing loss preferred to take their hearing aid out of their ear when listening to music. Now, they can turn on various settings through their phone to optimize the sound quality."

"Mom, I'm done," Bertie announced with finality.

"You can play some of your new games then," said Bess. "We'll join you in the living room in a little while."

"Oh! We forgot the cranberry baked apples!" Naomi sprang from her chair and headed into the kitchen.

"It wouldn't be Christmas if at least one thing hadn't gotten left in the oven," said Frank equably, opening another bottle of red wine. "Pass your glass, Theo."

When Naomi returned, Nan shared with her and Teddy some of her ideas for the wedding ceremony's music, and the talk at the table turned generally to the wedding in April. Naomi showed Bess and Nan a picture of the simple but classically styled dress she'd found at a vintage shop. Teddy relayed some of the responses he'd gotten from his e-invite, as well as calls he'd received from those friends and relations to whom he'd mailed out card invitations.

"We have at least seventeen anti-vaxxers coming," said Teddy. "They're all Auntie Jo's family. Emmaline's husband Chris emailed to say they'd 'do their best' to come with a Covid test result, and Curtis called to say they 'often forget their masks, but will try to remember to bring some.' I said we'll have plenty on hand for anyone who's not vaccinated, and he just changed the subject."

"I can see everyone is going to give Jo's family a wide berth at the wedding," said Frank dourly. "It's unfortunate that there are so many of them, it's going to be a challenge to steer clear of them all."

Bess sighed. "They have that unshakeable Evangelical belief that God will protect believers from illness."

Frank added, "And then when so many people die, it becomes part of the narrative that the end of the world is near. Or it's a punishment from God for the sins of some people."

Bess helped herself to a baked apple. "I just hope none of Jo's family starts preaching to people in the midst of the reception. That would be so tacky—especially if they're the only ones wearing masks."

"Maybe I should include that in the instructions over by where the masks and Covid tests are," said Teddy thoughtfully. "'No preaching if you're wearing a mask.'"

After they'd cleared the dishes away, they returned to the living room, where they could take a break from food for a few minutes. Bess resurrected the fire in the grate, and Nan was about to put on some Christmas carols, when Frank said, "Nan, how about you play us something festive?"

She was delighted to offer them some piano. She played the winter pieces of Tchaikovsky's *The Seasons*, and then, still in the mood for Tchaikovsky, she played some of *The Nutcracker Suite* as arranged for piano. After that, thinking of Thomas and Austria, and as a nod to the Vienna Philharmonic's New Year's festivities, she gave them some of Johann Strauss's waltzes. Elise invited Theo to dance and, with a good-natured smile, he stood up and fumbled through the moves with her. As they stepped about the room, the low hum of conversation among Rebecca, Naomi, Bess, Teddy, and Frank provided a calm, steady burden to the whirl of the waltz, and Nan saw the snow softly beginning to fall outside again.

Chapter Nineteen

The days leading up to New Year's Eve had lower skies and moister air, and the snow that came at least once each day bestowed a fresh layer of pure white on the mounting drifts. One afternoon Bertie and Nan built a snow Pooh-bah and made hot cocoa when they came in. One evening Bertie, Nan, and Bess roasted marshmallows over the fire in the kitchen stove. Another afternoon all three of them sledded from the top of the large backyard to the very bottom, perfecting their track each time they returned to do the course again.

Nan spent two hours at the piano each day working out the last remnants of her second movement and playing back over parts from her first and third movements. Upon Frank's request, she also played and sang Schubert songs that were in the mezzo soprano range—most notably the song cycle called *Die Winterreise*, or *A Winter Journey*, by Wilhelm Müller. When she sang the German lyrics, she thought of Thomas, and one song in particular brought him to mind. In German the title was "Mut," which translated as "courage." The lyrics went:

If the snow flies in my face
I shake it off.
When my heart in my bosom speaks,
I sing out loud and merrily.

I hear not what it tells me,
I have no ears for it;
I feel not its complaint,
Complaining is for fools.

Into the world then merrily,
Braving wind and weather!
If there is no God on earth
Then we ourselves are gods!

Nan held her final Meetup of the year on the 27th, in which at least half of the participants had accomplished what they'd aimed to achieve by the end of the year. Nan knew she was just about done, and she shared with them the part of the second movement that she'd been working on that day.

Margot called to say that Aaron wanted to get together on New Year's Eve. They were going to a Turkish restaurant in the City that would stay open till 3 a.m. She sounded glad to be seeing him again.

"What happened with James?" asked Nan.

"He was a darling, but I found I kept comparing him to Aaron and missing Aaron."

"You guys are going to get back together tomorrow night, I can feel it," said Nan excitedly.

"We'll see." Margot seemed equal parts subdued and expectant.

Nan was relieved that, like herself, Thomas didn't seem too keen to repeat their Facetime experience. She suspected it was another sign that they were both too much people of the mind to relish the shortcomings of the screen. She much preferred talking to him on the phone and texting, as they did several times over the course of the week.

"I fear I've lost you to the enchanting wastes of snowy Minnesota," said Thomas as they talked on the 30th.

"How so?"

"Your talk is full of the winter weather, outdoor activities, songs of the season, fire and ice, and you make it sound like paradise."

"It is," she said, "except for the fact that you're not here."

"What makes you so sure I belong in paradise?" he asked with his usual irony.

"Because I'm here," she said ambiguously.

"Nicely put."

"How did the grand opening go last night?"

"It was fraught. Some of the supplies didn't come in that Luca needed, so we ended up deleting a few of the menu items. One of the cooks got nervous under Luca's scrutiny and dropped a whole tray of fried polenta. We offered one free drink to everyone, from off of a menu of five drinks, and Luiz, Elian, and I never got a second's rest for all the orders people placed. Overall, I think the night was successful at getting the buzz going for The Chaser, and we got some good feedback on the bar bites."

"Will it be even crazier tomorrow night?"

"You know it. We're offering a New Year's Eve package of a bottle of champagne, a selection of bar bites, and the whole New Year's kit of hats, noise makers, and party horns. I drew the line at confetti, however. That stuff takes a long time to clean up."

"Will we have a chance to talk around 10 your time?"

"You mean midnight yours?" He had a smile in his voice. "Sure, I'll call you."

Nan wanted to say something more, but she let a pause ensue, and the moment passed. She suddenly felt shy again with Thomas.

"Talk to you then?" he said.

"Talk to you tomorrow evening."

When Thomas had rung off, Nan wondered if he too had felt tempted to say "I love you" during their call. They were in such a weird limbo right now, having been separated for six days directly after owning up to their feelings. But then she figured she was, as usual, overreacting by analyzing all of this so intensely.

Frank and Nan took a bracing six-mile walk in the snow to the lake and back on New Year's Eve afternoon.

"Thomas wants to go cycling together in Hungary and Austria in June." Nan longed to make him present, using words to do so.

"That'll be grand—you two seem well matched," said Frank with great warmth. "I look forward to meeting him. You are bringing him to Teddy's wedding, right?"

"I haven't asked him yet." Nan flushed. "But he might be able to get a few days off for it."

"If he loves you, which it sounds as if he does, he'll be able to come."

"Is love always filled with uncertainty?"

Frank didn't waver. "If it's worthy of the name of love, it is."

"You mean sort of like the unexamined life, the unexamined love isn't worth feeling?"

"No. Rather, desire rejects reason, which seeks to square things nicely."

"So it's the contest between desire and reason that makes for the uncertainty?"

"More or less."

Frank had placed a leg of lamb in the sous-vide to slow-cook all day, and when they got back from their walk, he directed Nan to bring out the ingredients for assembling the blini, crème fraiche, and caviar for their appetizers. He had three bottles of champagne chilling, and he and Nan began to prepare the roasted vegetables to go with the lamb.

"I should let you go finish your composition so we have one more thing to celebrate tonight," he said.

She laughed. "I finished it this morning, actually!"

"Oh! Then we'll get to start drinking a bit earlier."

As soon as Bess and Bertie came back from a trip to the mall and Teddy and Naomi arrived, Nan and Frank got the festivities underway. They put on the sophisticated classical radio station of the Twin Cities and everyone drank too much and ate too little, in anticipation of the main course. The lamb was tender and filled with flavors of rosemary and mint, and Nan helped herself to three portions, she was so enamored of its taste. By 11, even Teddy and Naomi had had a full glass of champagne each, which was exceptional for them.

"Now, I mustn't forget the crème brûlées," said Frank.

With great fanfare he produced his blowtorch and caramelized the sugar on the top of each crème brûlée. Bess was pleased to see that Bertie not only enjoyed the custardy texture of the dessert, but loved cracking its brittle surface.

"Can I crack yours too?" he asked Nan.

"Go ahead," she said. "I was needing someone to do my cracking."

At exactly 11:55, Thomas called. Nan broke out in a smile and moved to the other room, where she could hear him better.

"Sorry if it's tough to hear me—things are a little bit loud here," said Thomas.

"Good business?"

"I've never seen it like this. It's as if we're the new trendy spot in Uptown. We've got newcomers, regulars, Millennials, Boomers, gay, straight, affluent, bohemian, Black, white, brown, and even Asian— I've never been able to attract enough Asians!"

"Congratulations."

"Congratulations to you too, on finishing your composition." Thomas paused. "And . . . I believe it's Happy New Year's for you, as well."

Nan heard her family counting down in the kitchen and she said, "Happy new year."

"If I were there, you know what I'd be doing."

"What would you be doing?" Nan smiled.

"I'd have my hands on your butt and I'd pull you to me, and the kiss I'd give you would be censored."

"Why so?"

"My tongue would devour your mouth, and I'd make your lips swell up from biting and sucking them."

She laughed. "Can we do all of that three days from now?"

"Would you like to try the scarves the evening you come back?"

"Yes, let's."

When they'd rung off, Nan returned to the kitchen, where Bess said, "Is Thomas's place hopping?"

"He's making a mint on champagne. And everyone and their grandmother has showed up."

Frank refilled Nan's glass with champagne and they all clinked glasses and wished one another a Happy New Year's.

"And, lest we forget," added Frank, turning to Nan, "congratulations on finishing your quintet!"

"When will we get to hear it?" asked Bess.

"If I can't get anyone to play it before my own quintet does, then the Perseid quintet will play it in late May," said Nan. "In the meantime, I'm sending the draft off the SFSU people so they can get a sense of what it's like."

Bess sipped her champagne. "Do you feel good about the first draft?"

"Right now, I love it. But I'm going to leave it for two weeks to gain some perspective, and then come back to it in mid-January."

"What will you do in the interim?" asked Frank.

"Make a holiday watercolor card," Nan declared, having already decided the subject for her painting and its rough composition.

New Year's Day, Nan was completely absorbed in sketching out her holiday picture, which she invariably drew on a thick piece of 11 ½ x 17 watercolor paper. Inspired by her walks with Bertie and Frank, she sketched a group of dogs that looked just like Pooh-bah engaged in a snowball fight down by the creek. Some of them had ducked behind trees, others had built up mini-forts as defenses, and others were hurling snowballs over their shoulders as they ran. They'd made a cozy fire off to the side, where a lone Pooh-bah sat on a log warming his paws. The sun had just set behind the fir and spruce trees, and the lengthened shadows of the dogs in movement made up another character in the picture. Nan decided, as she began painting, that she was going to put her "Every dusk is a miracle" poem on the inside of the card.

The painting took the rest of her stay in Minneapolis to complete. Nan sat at the kitchen island listening to the classical radio station until

3 a.m. two nights in a row filling in her sketch with as vibrant and varied a series of colors as she could mix from her watercolor tubes. The process was deeply meditative in a way that poetry, cycling, and even music could never be. In the middle of the night before she was due to leave, she finally left the painting, and when she woke up the next morning, she touched it up a little and held it at a distance. It was done at last. She borrowed the Subaru and drove to a copy shop where she could scan the painting and then shrink it down to fit on a regular paper that had been folded in half. Then she bought a number of red and green papers, two glue sticks, and a bunch of large envelopes. All that morning she cut out the xeroxed picture and glued it onto the folded colored papers. By noon, she had finished this first step of assembling the cards. When she got back to Berkeley she would glue the poem on the inside of the cards, and write more individualized messages on the cards for all the people she was sending them to.

At the airport, she received a text from Margot: "Aaron wants to give our relationship another try. I said yes."

Nan texted back, "Hooray!! So happy for you, Marg. I can't wait to meet him."

Nan could never sleep on planes, but this time, on the flight going back to San Francisco, she was so exhausted that she somehow managed to doze off for almost an hour. When she awoke, the flight attendant was announcing their imminent landing. Nan realized she was starving, even as she remembered it was a mere half-hour before she'd be seeing Thomas again. The combination of hunger and excitement, coupled with her fatigue, kept her wired and keen. As she stood on the curbside keeping an eye out for Thomas's car, she felt a warm hand slip around her waist. Startled, she turned and met with Thomas's lips, as he tipped her face up to his. Their kiss was sensual and full of longing, and Nan never wanted it to end. She leaned into him and he brought her in even closer with his arms.

When their lips parted at last, she ventured to ask, "How did you come from that direction?"

"It looked as though your flight was going to be delayed at the last minute, so I did short-term parking. But then you were on time after all."

Thomas led her to his car and placed her suitcase in the trunk. Then he closed the distance between them and kissed her again, this time using his tongue to express things he hadn't said in their first kiss of reunion. Her entire body responded to his nearness, his kiss, the touch of his hands, his presence. She shivered with delight and felt pulsations between her thighs that grew in intensity as the kiss lingered on. Thomas was wearing a long black wool overcoat that made him look even more dashing than she'd conjured him in her mind's eye. She traced her fingers along the shadow of stubble that darkened his jaw line and was flooded with love and unconfined desire.

As Thomas drove, he kept his hand on Nan's thigh, and she placed her hand over his. They rode in complete silence. Every so often he looked over at her with a fiery gleam in his eyes, and then turned back to the road. It was the most tantalizing drive she'd ever experienced, for she fancied she knew what awaited them both when they got back to his place.

♪ ♪ ♪ ♪

When they entered his house, her stomach rumbled hollowly, reminding her how hungry she was. But when she turned around to suggest they eat something, she was struck speechless by the transformation in Thomas's demeanor. His expression had turned into that of a predator bent on cornering its prey, and she instantly found her breaths coming more quickly and shallowly in response. Although he hadn't as yet made a move, she backed up a few steps and stumbled into the back of an armchair. She didn't dare look away from his eyes, which had effectively arrested her in her tracks. Still wearing his long coat, he advanced smoothly towards where she gripped the chair back, came within half an inch of her so that she could feel his breath on her neck, and allowed his eyes to rove down to her bare collarbone as if considering biting it. No more distance separated them than separated a pair of tango dancers, and the effect was just as intimate and dangerous.

He turned over one of her hands on the chairback and lingeringly drew his two forefingers up the inside of her wrist. She shuddered involuntarily from the suggestion of what he was contemplating as he felt her wrist. She imagined he was calculating how he would fasten a silk scarf around it. Then, using the same fingers, he traced a subtle pattern over her lips before cradling her chin in the crook between his thumb and index finger. It was as if he was taking inventory of all the parts he would soon be handling as spoils of his conquest.

Nan had by now checked all thoughts of Thomas's tenderness, steadiness, and gentility; her brain had once more tuned in to that dark

304

side of him that resembled the devil—that was capable of cunning, tyranny, and even cruelty. Yet, as she adjusted her attitude from trust to wariness, she realized she loved this dark side as much as she loved the lighter side. In fact, she conceived of both as fitting her temperament the way tongued and grooved boards fit: where her mania made her impulsive, he remained steady; where she showed insecurity, he reassured her; where she used circumspection, he acted as a driving force to which she could give herself over confidently and wholeheartedly. The obverse of his presence of mind, however, was his cunning; the obverse of his assurance was his tyranny; and the obverse of his drivenness and forcefulness was his cruelty. As she looked into his darkened eyes, she was overcome with the conviction of her love.

He seemed taken over by a force greater than he. He brushed aside her hair, leaned in, kissed the base of her neck, and then gently bit the skin near where he had kissed her. She let out a low moan of surprise and delight. But when she attempted to kiss him in response, he gathered her hair in his hand and pulled firmly on it, tilting her head back and keeping it immobilized as he laid a few more prolonged kisses on her neck. With his other hand he opened up the front of her shirt and gripped her waist. As she tried to put her arms around him, he tugged her hair more vigorously, forcing her attention back to his kisses, which he followed with a few more bites and puffs of cool breath. His ruthlessness yielded equal parts pleasure and torment, which she willingly bore even as she suspected this was only the beginning of her agonies. When he released her at last, she prepared to meet his lips, but he took a step away, his eyebrows drawn intently and his eyes fixed on her broodingly.

In the wake of this first demonstration of his power, some part of her determined not to concede readily to him but, rather, to test his purpose and resolve. With the utmost effort she looked down from his enthralling gaze and shook her head slowly as if by doing so she could break the spell he had bound her with. She expected him to renew his claim on her, but he surprised her by stepping away and taking a seat on the couch. She turned around to follow his movements, and a flash of bright color drew her closer, her curiosity getting the better of her. He held a silk scarf that he was by turns smoothing out and twisting around his hand. This hypnotic motion riveted her to the spot, as she faced the couch about four feet away from where he sat.

A smile played about his lips as he extended the scarf between his two hands, pulled it taut, and then relaxed it. She willed him to speak, but he remained sphinx-like, his brows knitted even as he smiled. His face was a spring day in which showers and sun collided. Then she realized he was engaging her in a battle of wills to see who could remain silent the longest. If that was the challenge, she would rise to it.

He looked back up, released the scarf in one hand, and crooked his two forefingers at her. Had she not been focusing on his every movement, she might have missed the gesture, it was so subtle. Without thinking, she stepped forward. Too late she discovered her mistake: in one deft motion he bound her wrists together with the scarf and pulled her down beside him on the couch. He fastened the scarf so swiftly that she scarcely had time to register what was happening. He produced another scarf from his pocket and secured her ankles in another brisk series of movements. Strong as she was, Nan found she genuinely could not move her arms or legs in this position. She wriggled her hips, but stopped when she saw him smile.

He had placed a firm arm across her pelvic area so that she couldn't stand up, and with the fingers of his other hand he traced increasingly small circles around and over her ear so that at last his middle finger was on the threshold of her ear canal. The language of his gestures was unmistakable: he was taking possession and advantage of her at once. He settled in closer and looked into her eyes. She read in his a playful but provocative dare, and her own conveyed her desire to continue the contest.

Then he approached her where she was most vulnerable. He brought his lips to the verge of her mouth and hovered there for a long moment. His lips looked shapely and tempting and it was all she could do not to close the gap by leaning forward. He smiled then, seeming to guess at her impulse as well as her restraint. Finally, not believing he could be so merciless as to continue to deny her, she reached her lips longingly towards his. Just as they nearly made contact, he withdrew. Frustrated, she decided to test the force of the arm that secured her hips. She attempted to lift up her legs together with her shackled hands. With ease he kept her pinned in place, while using the fingers of his other hand to trace a proprietary line from the top of her neck down, skirting her breast, trailing over her ribcage, and pausing at her waist.

She trembled beneath his touch, even as she began to weaken in her resolve. His methods of torture were so perfectly calculated to drive her to distraction, that her mind had little room left to contemplate resistance. Yet with great effort, she determined to try one last thing. She twisted her torso so that it was flush with his, and so that the muscular arm that hemmed her in was now locking in place her hip rather than her pelvis. She immediately felt the advantage this gave her in maneuvering as well as in strength. But Thomas, divining her thoughts, pulled her to him and brought all plans of escape to an end with a kiss. She marvelled at the assurance with which he quelled her struggle—a small part of her even resented him for his ability to govern her so completely. But her body and mind were both clamoring to have him subdue them, and at last she gave in to their desires. She closed her eyes and groaned deeply, allowing his tongue to guide hers in their dance.

The urge to compete gave way entirely to the flood of desire and love that overwhelmed Nan at that moment. She reveled in Thomas's superior force, even as she savored her own utter helplessness.

She was on fire and could barely stand the time it took Thomas to unfasten her ankles so that she could move to the bedroom. He kept her reined in, proceeding at the pace of a snail, as he shadowed her there, pulling her to him from behind each time she advanced too hastily. Every so often he would remind her of his power by cradling her neck in one hand and placing his other hand between her legs. Once they had reached the bedroom, he closed the door, untied her wrists, and stood back, still in his coat, observing her from the distance of six feet.

At last he spoke. "Take off your jacket."

She did as he ordered.

His voice became lower. "Now take off your shirt."

She did so.

Now his voice was gravelly. "Remove your jeans."

She obeyed.

He paused for a long moment as if this was his favorite part. "Take off your bra."

She followed his command. A long pause ensued, as he drank her in. "Take off your underwear."

She did so.

At last she was entirely naked in front of him, and she wondered if he could tell how urgently she desired him. He circled slowly around her much as she imagined a hawk would do when closing in on its prey. Finally, he stood directly behind her and moved his hands from her breasts down to her groin, pulling her tightly to him.

He whispered into her ear in a way that made her quiver all over, "Now, lie down on your back."

She thought she couldn't be more turned on than she was now. But what followed drove her more insane than her wildest imaginings. After she'd lain back on the bed, he secured a silk scarf to each of her wrists and ankles and tied up their ends to the bedposts. Splayed out and completely exposed, she felt even more powerless than she had when he had her pinioned on the couch. She closed her eyes and, once again, images swiftly played through her mind of Thomas encircling her in his dark coat as if preparing to pounce, of him teasing her by removing his lips just out of reach, of him tracing her wrist with his fingers. And then his real presence hovered over her, he at last naked like her, and she longed to have him inside her. She lifted her hips as much as the scarves would allow and twisted in anticipation of his entry. He teased her for a while with his fingers, lips, and tongue, until she could stand it no longer. By the time he touched and entered her, she shattered in a million pieces, and he followed soon after. It was the most in-the-body experience Nan had ever had, and she continued to come as Thomas remained in her.

Her first reflection after coming apart was that Thomas played her with the same mastery, the same passionate intensity, that she played the piano. The expression he'd assumed at the start of their seduction today was the look of an artist in the grip of inspiration.

After he had released her from the scarves, and as he lay next to her, he murmured, "That has been a long time in coming."

"You mean because you conceived of it as soon as you met me?"

He shook his head. "The bondage is of secondary importance. I meant the fruits of the long seduction. It feels as if this process keeps getting more and more beautiful, and today was the culmination of all the trust and knowledge of each other that we've built up over a long time."

"I wonder if it's possible to keep everything just as exciting and uncertain as it was just now."

"As long as we both continue to test each other and engage in conflict, anything is possible."

"But will you always win?"

Thomas kissed her lingeringly. "Let that remain in question, if you will."

Chapter Twenty
Thomas

Nan stayed at Thomas's place for six nights, and in that time, they made tender love, fierce love, tense love, reckless love, needy love, and propitiatory love as the occasion dictated. Nan kept saying she ought to go back to her place for a night, but Thomas invariably persuaded her to remain another day. He was glad to have confirmation, in one area of life at least, that he retained control over what mattered. In making love to Nan, his perfectionism almost always paid off. These days, such satisfaction was rare. The rain had poured steadily while Nan had been gone, and he had foregone outdoor riding—had opted instead to ride on the rollers in his front office, an activity which had served as a poor substitute for the fresh air and hills. Consequently, his cycling fitness had suffered. Further, since business had been booming at the bar, he had missed his Meetups for oral language practice and had been too tired to do much reading in any of his languages. His brief trip to Baltimore had taxed him more than he cared to admit, and he had missed Nan much more than he thought possible. He began to suspect that exhaustion from the past month was taking its toll on him.

Since the new year, he had made several critical mistakes at The Chaser that he knew stemmed from driving himself too hard. An online magazine had done a profile of the place and quoted him out of context as he had commented on the role he hoped The Chaser would have in Uptown. As a result of the article, it sounded as though Thomas was criticizing other restaurants in the area and trumpeting his own place as filling a much-needed gap. He'd tried to have the writer correct this, but she'd claimed it was too late and stuck by the original version. The other evening, Thomas had supported one of the sous-chefs with whom Luca had picked a fight, and Luca had stormed off, returning peaceably two hours later. The incident had confirmed Thomas's suspicions that however temperamental Luca was, he blew off his anger relatively quickly and held no grudges. Several times now, employees had complained about misunderstandings related to time off, staffing, and overtime shifts. Since Thomas was certain that much of this confusion came from his own errors, and that he was stretched to the breaking point, he had decided to hire a restaurant-bar manager to deal with these matters. Meanwhile, a few of the new staff had tried to interfere with the music, and Thomas had clumsily refereed a squabble between Luiz and Elian, on the one hand, and the newly-employed adherents of pop, electronic, and rap, on the other hand. Thomas was not averse to introducing some new kinds of music, but he trusted Elian's judgement when Elian pointed out that it made sense to time the music according to the clientele who was most likely to enjoy it. Thomas had even lost his temper and snapped at one of the servers who was slouchily watching videos on his phone one afternoon when they were preparing for a busy evening. Thomas's biting comment had roused the server from his stupor, but had caused the rest of the staff to raise their eyebrows in surprise.

During this time of duress, Thomas allowed Nan to see him at his most controlling, even as a part of him wished he could temper this tendency. She had come into the restaurant one night to help out, and a patron with whom Thomas was well acquainted had requested a drink that Thomas told Nan required a goblet glass. While Thomas genially conversed with the patron, to whom Thomas had passed the bottle, Nan looked around for the glass and came up with what she thought was a good fit. She brought out a grappa glass she'd found, and the patron poured his drink into it. Thomas drew Nan aside and said, "But there should have been a goblet glass behind the grappa glasses." He then took her to the shelves, confirmed that indeed there were a few goblet glasses left, and explained the distinctions among the various glasses. Nan cheerfully took it all as a learning experience.

On another occasion, while making dinner for the two of them, Nan had cut onions on a chopping board Thomas reserved for vegetables that were not in the onion or garlic family. Thomas had made a joke of it, saying you can never have enough onion taste in anything, but Nan must have noticed his tenseness, for she came over and softly massaged his trapezius muscles. Nan didn't seem particularly to care how she washed her vegetables—or if she washed them at all—so Thomas painstakingly showed her how to soak the Romaine lettuce leaves, separate them, and towel-dry them afterwards. He also demonstrated how he soaked mushrooms for twenty minutes before cutting them up. He was alarmed to discover how out of sorts his pans quickly became after she used them and replaced them wrongly on their various hooks. He just barely managed to stop her from using an abrasive scouring pad on one of his best frying pans. He hoped Nan didn't notice how precise his language became as he explained everything to her, or how edgy he was while he did this. One day, before they'd headed out on a bike ride, Nan had applied to her bike chain an expensive mechanic's lube that Thomas scrupulously reserved for his cables, and he'd had to spell out the difference between these lubes. On another day, Nan had started a laundry in which she'd combined their athletic clothes with towels and other dirty linen, and after Thomas had discovered this, he'd stopped the machine and separated out their sweaty gear from the rest of the laundry.

Despite all of these skirmishes, Thomas found himself unprepared for the evening, at the end of the week, when Nan announced she finally had to go home to take care of some pressing practical matters. To Thomas, those twenty-four hours without Nan seemed endless, as he used them to deal with bills and accounts, details for the full menu that Luca had rolled out the previous week, schedules for interviews with more food reviewers, and health protocols he and Luca were reinforcing in anticipation of an imminent food inspection. He realized he had gone into the bar-restaurant every day for the past twelve days for at least ten hours each day, and found he was beginning to lose some of the enthusiasm he'd had for The Chaser towards the end of 2021. Was he getting burnout already? He hoped not. After all, this was early innings.

Thomas was vastly relieved when Nan called, that Monday evening, to ask if she could come back to his place. It turned out she hadn't gotten the SFSU job. She'd received an email from the director of the music school that morning saying they'd had a lot of well-qualified applicants and had decided to go with another candidate. When she arrived at his house, she told Thomas she was grateful that they'd let her know one way or another, as so often in academic searches one either never heard from the school again, or one had to tease out a definitive response after much waiting and suspense. But he began to notice her sink into a depression as one after another of her many attempts at finding a permanent job dissolved into nothing. On Thursday evening she burst into tears, and when Thomas put his arms around her, she explained that this was just the prelude to her period. But he could tell that she was also demoralized by the prolonged silence that greeted so many of her applications for full-time jobs.

Thomas's instinct, as always, was to troubleshoot.

"What have you put up on your YouTube channel so far?" he asked.

Although Nan's vitality had ebbed considerably, she brightened at the mention of the channel.

"Just the 'Ghost' trio from December. It's gotten a fair number of likes."

"And who's going to perform your quintet?"

"I need to contact that group in San Francisco to see if they can do it in the next month. I don't want to wait till May for my own quintet to tackle it."

"And as soon as you can get them to record it, you can add that to your YouTube channel?"

"I guess so."

"What if you dial back your search for full-time jobs for now, and try to attract more piano students—what would that require?"

Nan considered this. "I think if I offered the entire weekend for lessons, that would increase my numbers, and the students would show up a lot more consistently."

"So could you shift the bike riding to, say, Monday and Friday?"

"Would that work for you?"

Thomas squeezed her more tightly. "Friday is just as good as Saturday for me to ride, and Monday, as you know, I usually take off—so, yes."

"I'm also missing the composing. It was so all-absorbing, and so satisfying. I have nothing to replace it with."

"Can't you begin a new composition?"

Nan nodded. "I've already begun taking notes for a song cycle in French for mezzo soprano and piano—I hope to have Cécile perform it during the summer before she heads off to college."

Thomas kissed her cheek. "That sounds like a plan that's sure to succeed."

Nan laughed, and Thomas realized only then how gratified he was to hear the sound once more. "In the midst of all you have to worry about, how can you think so much of me?"

Thomas smiled. "Believe me, it's a relief to think about your problems instead of mine."

He didn't add that he was prouder of her than he'd ever been of any girlfriend, and that he derived great pleasure from helping her to advance in her career. Since he'd been sixteen, Thomas had never had more than a few months of his life go by in which he hadn't dated or been in a serious relationship with a woman. He and Nan had mercifully decided, by tacit agreement, not to talk about their past relationships, and Thomas rarely found himself comparing Nan to the many women he'd dated before her. She was simply in a league of her own. But he couldn't help feeling that all of his intimate experiences until now had merely prepared him for being with Nan, and he took each of their exchanges and encounters as another chance to learn about himself, about her, and about them together as a couple. Thus while he readily played the role of assured lover, he also acted as a porous sponge soaking up every detail that could enhance the fine art of seduction to which he devoted so much thought. He often compared the ephemeral arts of lovemaking and hospitality that he practiced to the longer-lasting arts that Nan cultivated, and his admiration of her deepened.

Thomas saw in Nan's drivenness the reflection of his own pursuit of perfection in fitness, languages, hospitality, and seduction. He acknowledged how inconsistent it was in him to love Nan's imperfections, and yet they were what he invariably fastened onto when he daydreamed or fantasized about her. He loved her insouciant impracticality in all the areas he controlled so carefully, her impecuniousness, her insecurity, her tunnel vision, and her high-strung nature. He suspected he cherished these faults because he could so easily cover for them with his own strengths—his exacting precision, his financial canniness, his confidence, his awareness of the total surroundings, and his composure. Yet what was love but selfishness in the end, however much it might afford an expansion of the self? He staunchly resisted the idea that loving a woman could be a selfless act.

Two weeks after Nan had returned, on Martin Luther King Day, as they rode together up Spruce on their way to climb Happy Valley, Nan said, "Now that your new restaurant-bar manager is settled in and doing well, what if you took a few days off?"

Thomas had briefly considered this once or twice and then dismissed it as too risky. "I'm not sure Craig is up for dealing with everything just yet."

"But surely Elian and Luiz will help him," Nan persisted. "You could check in with Craig every evening to make sure things are going smoothly and to make any critical decisions."

"To be honest, I'm not sure I wouldn't spend the whole time worrying so much, that it wouldn't negate the purpose of taking the time off," he said, with a rueful smile.

"What if we went away somewhere for a few days? Would that help take your mind off the bar and allow you to relax?"

Thomas recognized that in Nan's depressed state she would benefit as much as if not more than he from taking a road trip somewhere.

"Where did you have in mind?" he asked.

"My hiking friends love Kings Canyon National Park in the warmer months—though they say it's also lovely in winter. I've never been before—have you?"

"Near Mount Whitney? No, I haven't." Thomas reflected, swiftly weighing the pros and cons. The main con was the uncertainty of trusting everything to run smoothly in his absence at The Chaser. The main pro was that he and Nan could recharge while doing something new together—trekking in the snow—in an environment far removed from the everyday grind. "What about sleeping arrangements?"

Beneath Nan's attempt to sound casual, Thomas heard the eagerness. "My hiking buddies have recommended a lodge that's on the border between Kings Canyon and Sequoia parks. That way you and I can choose some day-long snowshoeing and cross-country skiing routes and have a warm shower and bed waiting afterwards."

Thomas grinned. "You've already done some research, I see. That's about a four hours' drive away, I'm guessing?"

Nan nodded.

"I could probably take off this Thursday morning through next Monday night. What about your students—and Albert?"

"Albert and I have shifted to meeting Tuesdays now. I'm in the process of rescheduling all my weekday students to Saturdays anyways, and my Sunday students won't mind holding off until the following Sunday to meet."

"Let's book it when we get back," said Thomas riding over next to Nan and wrapping his arm confidently around her back.

Nan's laugh was full of admiration. "Now you're just showing off!"

♪ ♪ ♪ ♪

As they drove through the Central Valley along Highway 99 on Thursday morning, Thomas felt as if he was playing hooky. Nan had used Bluetooth on her phone to connect to the car's sound system and was playing early rock and blues from the 1940s and 50s on Spotify. A steady stream of songs both familiar and period-specific provided a rhythmic backdrop to the fertile farmlands, orchards, and vineyards

314

they passed, even as the increasingly clear, imposing, snow-covered Sierra Nevada mountains defined the horizon to their left. Thomas had gone to Yosemite twice before, but he hadn't explored the Sierras to the south as much as he'd have liked. One summer nine years ago, when his friend Brad had visited from Colorado, the two of them had driven down the eastern side of the Sierras on route 395 and had enjoyed breathtaking views first of the mountains and then of the desert as they headed through Death Valley towards Las Vegas. Now, Thomas inhaled deeply, as though to capture the more rarified air he remembered from that trip at 8,000 feet almost a decade ago.

Around 11:30, just outside of Fresno, they pulled into a barbecue lunch spot that Nan had found on Yelp.

As they wolfed down some delicious ribs, brisket, macaroni and cheese, and cornbread for their first meal of the day, Nan said, "Do you ever feel as if the more you discover of California, the humbler you become?"

Thomas took a long pull of his coffee. "I can tell you haven't yet seen Yosemite. You wouldn't even ask that question after going there."

Nan gave him a gentle dig in the upper arm. "Way to rub in my greenness."

He laughed. "We'll go there someday. In the meantime, we've got your Kings Canyon itinerary to execute."

"There isn't really any itinerary, other than a few possible routes for skiing and snowshoeing, and enjoying the lodge."

Thomas had brought whiskey and lemons for whiskey sours, and he stopped at a gas station to fill the tank and pick up two bags of ice before they headed east on Route 180. Nan turned off the music, for the scenery rapidly became too stunning for either of them to process in conjunction with sound. The towering brilliant-red sequoias contrasted exquisitely with the thick white snow as they slowly climbed up more than 7,000 feet over the course of 39 miles. It was fortunate that today the park was not requiring four-wheel-drive vehicles to have chains on their tires, although Thomas had brought some just in case. They pulled into the lodge around 2 p.m., checked in, and carried their bags into their cabin, the deck of which looked out on a vast grove of sequoias, pines, and firs and, in the distance, the snow-capped peaks of Sequoia National Park.

The temperature, in the low 20s, refreshed and exhilarated Thomas.

"Did you say they had a hot tub here?" he asked, pulling Nan in for a kiss.

"Yes. Somewhere down there, I think." She gestured down the slope from their cabin.

"Let's go find it."

Nan brought her bikini and Thomas his swim trunks, and, after asking directions from a couple who were taking off their snow boots outside a cabin below them, they wended their way to the hot tub. Luckily, it was an unpopular hour for congregating there—everyone else seemed to be out enjoying the snow. After they'd changed, they dipped their toes into the water, and just as hastily took them out again.

"Bess has an infallible method for getting into scorching hot tubs," said Nan. "First, you stand briefly under a lukewarm shower. Then you climb into the hot tub and dunk yourself, all except for your feet, for no more than five seconds—and then get out of the tub again. After a minute, you climb into the tub again and keep your feet elevated for the entire time you're in it."

Thomas nodded. "I think that's somewhat akin to how the polar bear swimmers tackle cold water. Let's try it."

A few minutes later, they were facing one another with their feet resting on the rim of the tub, immersed in the scalding water. Thomas admired the way Nan's wet hair, plastered over her scalp, accentuated the heart shape of her face, her deep blue eyes, and her swanlike neck. Submerging his entire body in the water, he moved closer to her and slid his hands up the outside of her arms to the base of her head, as he leaned in and kissed her. She closed her eyes and responded passionately with her tongue and lips, as she wrapped her arms around his back, and he felt her lush breasts press into his own hard chest. He reveled in her smooth skin and her curves, and he positioned himself so that his rise found a haven between her thighs. She leaned into him more deeply then and let out a sigh as they allowed their tongues to stroke and caress one another.

"This must be it," said a loud and merry voice from nearby. "I can feel the heat."

Thomas and Nan parted as suddenly as if they'd received an electric shock. The voice came from a portly woman in her late fifties who was leading a slightly larger man towards the hot tub. The two of them nodded affably to Thomas and Nan, and the man commented, "A great day for it, though, with these frigid temps."

The couple was visiting from Eugene, Oregon, and the woman, Anna, said this trip was a gift from their daughter for their wedding anniversary. Thomas was grateful that Nan engaged good-naturedly with them, for he wasn't currently in the mood to be sociable. After they'd soaked for another half-hour, he took the opportunity of a pause in the conversation to say to Nan, "Did you want to explore the grounds a little?"

Nan nodded. "I'm ready to get cold again."

Anna's husband chuckled. "If the heater in your cabin is anything like ours, you'll need to layer up a lot, even inside."

"Oh, is it broken?" asked Nan.

Anna explained. "The heaters here seem to be very weak. At least there's a fireplace in the main dining room."

Sure enough, when Thomas and Nan returned to their cabin, they found that the heater maxed out at a very low setting, and they both shivered a bit as the sun got lower in the sky.

"I know one way we can get warm again," he said, removing her jacket and kissing her neck. "We have unfinished business to take care of anyway."

"But I thought you wanted to explore the grounds," she reminded him playfully.

"I have other things I'd rather explore right now."

Awhile later, as they lay in bed, Thomas mused, "I can see that staying here will either make you very fit or very lazy."

"How so?"

"Since bed is the only warm place in this cabin, you either stay in bed, or you pile on your outdoor gear and venture out into the snow."

As if to compensate for the poor heating, the lodge had provided them with a copious supply of down comforters and blankets. Nevertheless, Nan had snuggled up as closely as she could to Thomas's body, and he found himself giving silent thanks for the faulty heater.

"Would you rather cross-country ski tomorrow, or snowshoe?" she asked.

"I have the feeling no matter which one we pick we'll be regretting it the day after tomorrow."

"You think we'll be sore, even though we train as hard as we do?"

"Skiing is no joke. And snowshoeing looks easier than it is."

After they'd had a couple of Thomas's potent whiskey sours, they headed over to the dining room.

"I don't understand why I'm hungry already," she puzzled. "You were the one who drove here, and I haven't done any exercise today."

"It's the mountain air," he suggested. "It gives you an appetite and makes you sleep more deeply."

The food was laughably mediocre, and Thomas caught a matching twinkle in Nan's eye as he put down his fork after a few bites.

"Have you ever eaten at a British bed and breakfast?" she asked mischievously.

"No, but I think I can guess why you bring it up now."

"I used to think they made the food bad on purpose so that the company shone that much more brightly."

"It's true that insipid food does make you seek solace in conversation. But we seem to have come on the earlier side—people are only now beginning to trickle in."

Just at that moment a family of four joined them, and the father, Hugh, told Nan and Thomas all about the nearby snowshoe and ski trails they'd tried out over the past few days.

"We didn't have time to venture all the way out the Panoramic Point trail today, but you two could probably manage it in a few hours, no problem," said Hugh. "It's supposed to have the most amazing views of Kings Canyon at the top."

His wife Julie chimed in. "For cross-country skiing, I highly recommend a trail that's just down this Generals Highway a bit, called Upper Woodward Winter Trail."

"Yeah, the skis they rent out here at this lodge are actually pretty good," Hugh said. "And if you're looking for a really challenging ski trail, try the Quail Flat Winter Trail up the road. You can make your route as long as you want."

"Doesn't this lodge have some phenomenal trails too?" asked Nan.

Julie nodded. "We weren't able to go to Big Baldy Peak, but you guys should try for it."

As Hugh and Julie turned their attention to their teen-aged son and daughter, Nan said to Thomas, "Would you like to start by snowshoeing out to Panoramic Point tomorrow and seeing Grant Grove while we're up in that area?"

"That's probably a good idea, since the road doesn't require chains just yet—but we can't bank on a storm holding off through the entire weekend." Thomas didn't add that he'd never actually put chains on his tires before, and so would have to YouTube it or ask someone to show him how.

After dinner, Thomas texted Craig to see how things were going at The Chaser. He received a reply back a few minutes later saying that it was especially busy for a Thursday night, and there was a massive birthday party going on. So many people had ordered the salmon that they had run out of it, and there was an inexplicable craze for tequila.

Thomas appreciated Craig's attention to detail and his skill at delegating tasks to people in a fair and efficient way. He seemed to be well liked by Elian, Luiz, and even Luca, and he was equally good with the wait staff. Thomas relaxed a little more into the weekend.

When he and Nan returned to their cabin, it felt like the most natural thing in the world to back her up to the bed, pull her thighs around his legs, and drop her down onto the soft surface below. She always seemed surprised that Thomas could handle her in this easy way—as though her body were lighter or he was stronger than she expected. He lowered himself over her and ran his fingers slowly down to her groin, tracing patterns over the area while he brushed his tongue over her lips, nipping and sucking them. With his other hand he caressed her breast until she let out a moan that hardened him almost instantly. Then he wanted their clothes off as quickly as possible. In one fierce motion he pulled her leggings and underwear down and made contact with her wetness, even as she lifted up and tugged his shirt off. He eased her back down with one hand while loosening his belt and pants with the other, casting them and his briefs aside. In another swift movement he took off her shirt and unclasped her bra.

Then, as always, he took a moment to feast on the sight of her naked and helpless before him, longing to be possessed and overcome by him. To torture them both, he usually prolonged this moment as much as he could stand to, while teasing her with every sensation he knew drove her most wild—his lips, his tongue, his breath, his touch, his tip. Often he would merely lay a finger on her clit and send her into quakings and convulsions that were the hottest thing he'd ever witnessed. Sometimes he would wait until he had tormented her to the edge of reason and forced her to recognize his total power over her before he would give her what she sought. At other times, he would begin the foreplay by deliberately awakening her competitive impulse as he challenged her to a battle of wills; then he delighted in making her resent him for winning with ease, even as her struggle lent piquancy to his conquest.

Now he dipped his fingers inside her and tasted her salty sweetness before he slowly entered her. His next thrust caused her to inhale sharply and dig her hands into the duvet. When she tried to wrap her arms around his back, he pinned her wrists above her head. He could feel her approaching the brink, and he moved the forefingers of his other hand around her clit in the way he knew she craved, as he continued to move in and out. Then he lowered his head and kissed her, and they both combusted at once, crying out with unparalleled abandon. Thomas later felt as if he had blacked out for the entire time he came with Nan. He was only vaguely aware, as she clenched him, that she came several more times, and he realized she had pushed free from the hand that no longer immobilized her arms with conviction. She had pulled him down over her and he had allowed his weight to crush her completely.

♪ ♪ ♪ ♪

The next day, after a largely silent two-hour snowshoe trek up to Panoramic Point, they reached the lookout around noon, taking in the majestic beauty of the canyons and, beyond them, the unfathomably high peaks of the Eastern Sierras. Thomas knew that Nan's quiet stemmed from awe, as did his own. She put her arm around him as together they read the placard that told of the geological origins for the mountains and canyons and provided a key to the natural landmarks. They took a number of pictures, including a few selfies in front of each major view. Thomas's own favorite view was of Hume Lake to the Northeast, which was framed by mountains that ranged in height from 9,000 to over 14,000 feet.

"I see now why California has such a rich tradition of naturalist writing," Nan reflected. "The wilderness isn't just one facet of living here—it's the main facet."

"We're very lucky to have had some far-seeing conservationists like John Muir and David Brower." Thomas pulled his glove on more tightly. "Without them, we wouldn't now be seeing or enjoying these monuments to survival and the environment."

She nodded, but her focus remained on the question of literary perspective. "I think of all the stories I enjoyed while growing up—by Mark Twain, Wallace Stegner, Jack London, John Steinbeck—and realize that I will never be able to separate my early ideas about California, as shaped by these authors, from the reality I see before me as I travel through it."

"Don't you think that their minds have largely created that reality, and that our minds, interpreting their works, have aided the process?" Thomas stepped up to the edge of the lookout. "I compare it to how visitors to Weimar see everything through the lens of Goethe, Herder, and Schiller."

Nan followed him to the rim. "It must be so. I think my reality also springs from all the stories my grandparents, dad, and uncle have told over the years about their time growing up in northern California, their work experiences, and the history of their parents' families settling here. I begin to realize it was the magnetic pull of these layered perspectives that drew me most strongly to the West when I was in New York."

When they got back to Grant Grove Village, they took off their snowshoes and walked in their boots out to Grant Grove, for the path had been cleared of snow. They made their way through the center of the Fallen Monarch tree as they headed to the General Grant, the second largest tree in the world. (The largest tree, the General Sherman, was currently inaccessible because of road closures to the south of where they were staying on Generals Highway.) Both of them quickly gave up trying to capture in photos the magnitude and awesomeness of the 3,000-year-old sequoia; rather, they turned around and around gazing up towards the canopy of leaves far above them with the sky just beyond. Thomas calculated that seven people could lie head to toe at the base of the trunk and only just fill out its diameter.

When they had returned to the car, Thomas heard Nan's stomach growl and his own rumbled in solidarity. "Maybe we should pick up a snack here in the Village, since dinner won't be for another few hours," he proposed.

They bought some cheese and bread and a bottle of red Paso Robles Zinfandel and drove slowly back to the lodge.

"Oh, how I long for that hot tub again—do you think it'll be crowded now?" asked Nan when they got into their cabin.

Thomas laughed in disbelief. "Are you choosing the hot tub over food?"

In the end, they decided to let the popularity of the tub decide the matter for them. Since it proved to be empty, they repeated their ritual of the previous day and were able to soak and make out in peace for as long as their bodies could stand the heat—or either of them could stand the sexual tension.

"We seem to be putting off our hunger as we satisfy yet one more need," said Nan, laughing as Thomas impatiently tore off her clothes again after they'd returned to their cabin.

"This is going to be very quick," he said in a business-like tone.

"What about the art of it, though?" she teased. "That can't be rushed."

But Thomas made sure she regretted teasing him, for he turned her over onto her belly on the bed and aroused her first by lightly running his fingers a few times from her groin to her backside. He fastened a silk scarf around her wrists behind her back, and then tied a dark scarf over her eyes so that she was blindfolded. He looked longingly at her neck, wishing he had a collar for it. Once he had her subjected, he brushed back her hair and hovered over her ear, whispering, "Is there anything else you'd like to say?" He held his hand over the back of her neck gently but suggestively, as she cried out, "Please, take me, Thomas."

"I thought you didn't want this to be quick." He tickled her perineum with his finger. "I can draw this out until dinner."

"No, no! I need you now."

"I'm enjoying the rear view of you. You're so shapely and firm." He lovingly handled her muscular buttocks, large hips, and narrow waist for a few leisurely moments, while continuing to trace a line up her tailbone with his length. She raised her backside in longing and he parted her thighs expertly. Then she gave an agonized moan and he leaned in once more over her ears, murmuring, "I'm willing to be merciful. After all, you *are* made for love; you can't help fascinating me as you do."

"Oh, oh, *please*." She tried to turn her head over to the other side, but he pulled her hair assertively, keeping her facing the same direction.

"I'm going to give you a kiss, and then I'll give you a little of my hardness."

She wriggled with eagerness at these promises, and Thomas added, with a touch of cruelty, "But first, I want to hear you admit defeat."

"You have absolute power over me. You own me. I can't live without you."

"And remember, this is my game. I decide what we play, when we play it, how we play it. The rules change according to my whims."

"Yes, yes. Always."

He tugged her hair one last time before releasing her. When he kissed her he found himself losing a good deal of his restraint in the kiss. She was so alluring and voluptuous that he wanted to sink himself deeply in her and never emerge. But he had trained himself to discipline in this, as in other fields, and he fought to preserve his self-control as he proceeded slowly to plunge into her softness, while sweeping his tongue masterfully through her mouth. This rear-oriented position of dominance seemed to accelerate the frenzy for both of them. Nevertheless, when he touched her clit, he was only partially prepared for the detonation that occurred as if he had set off dynamite: they both disintegrated at the same time.

In the dining room, tonight's food tasted three times as good as the night before.

"I begin to see how I could cut some corners in the restaurant," he remarked as they ate their salad. "If I required all my customers to exercise in the snow the entire day before coming to The Chaser, I'd have a steady stream of rave reviews on Yelp."

"It helps to be the only place to which your patrons can resort after their exertions," she observed.

He smiled wickedly. "If you mean that as an allegory for our own lovemaking, I hope you realize one can construe it in two ways."

She rejoined, "Just as I assume you've worked out these mundane considerations in your business, I take it as a given that our lovemaking will eventually become pedestrian for you."

His eyebrows shot up. "And what gives you warrant for thinking as much?"

"You've said several times that for you, love is entirely selfish." She calmly took a sip of her wine. "Yet you acknowledge that the self is not everything. Therefore, in your view, love is severely limited. And where love is limited, lovemaking can scarcely hold interest."

He was momentarily nonplussed, and he bought time with an idle question. "Is this about my having teased you earlier?"

She shook her head dismissively. "You know I love everything we do in that arena. I only question how invested you are beyond the 'self-interested motives' you mentioned on our first date."

He collected himself for debate, deciding to attack her where she was most liable to appreciate it. "Let's take the other extreme. The ideal of a selfless love. When has it ever proven to be anything less than sexual and, therefore, self-interested? Abelard and Eloise, Tristan and Isolde, the troubadours . . . the forbidden love was always based on the erotic thunderbolt that shot through the eyes or ears to the heart."

"Based on your interpretation of these examples, you would sooner lay your life down for your good friend than for me," she said, refusing to rise to the bait. "For your claim that erotic love must have a selfish basis; whereas friendship, I take it, you still allow to be selfless."

Once more, he was struck mute. He had, indeed, had two or three extremely close friends from his high school years through college, notably Yevgeny, for whom he suspected he would have laid down his life. Did he place Nan in a lesser category than them?

She seemed to read his thoughts. "'Such is my love—to thee I so belong— That for thy right myself will bear all wrong.'"

Recognizing the final couplet of one of Shakespeare's sonnets, Thomas laughed uneasily. "Even for his friend Shakespeare's poet very seldom shows selfless love. By his own admission, he only stands to gain by writing about him. Though I suppose in a couple of instances, he does slander himself on his friend's behalf."

"And what about sonnet 116?" she countered. "The main idea is that true and lasting love has no ulterior motives."

"It's just that—an idea. I'll grant that you can occasionally show selfless love for your children, for your parents, and maybe even for your closest friends, as you suggested. But erotic love taints things and makes a person's motives questionable."

"I've often wondered if you really mean what you say." She looked askance at him. "Have you never had a girlfriend whom you also considered your best friend?"

"Once, yes. But since I never got a chance to prove my selflessness before I desired her sexually, I'll never know if that was a selfless love or not."

"Can you imagine yourself risking your own well-being for her in the absence of sex?" He noticed that she looked down as she said this, and he sensed she was making an effort to keep the hypothetical question impersonal.

He considered this. "I suspect you can paint any number of fanciful scenarios in which you demonstrate heroism for the object of your passion. But to me, this self-flattery only confirms the egotism of our motives once sex is involved."

"Well, since we disagree over the premise, and I can't change your mind over that premise, I guess we have a standoff," she said resignedly. "Centuries of philosophers have been unable to disprove your system of belief—I don't see how I could do so in one evening."

He smiled. "Well, whether I believe in selfless love or not, it shouldn't make much of a difference on our love."

She put down her wine glass and directed at him a look in which he was shocked to perceive a trace of disdain. "Only that it makes what we have no better than an infatuation, and it makes it all the easier for us to take one another for granted."

He reeled momentarily at her suggestion and, once again, he retreated behind humor. "I think you exaggerate a bit here. If anything, you've just uttered a paradox: it takes time for a couple to grow to the point of taking one another for granted; an infatuation is a short-term affair."

She shook her head impatiently, taking umbrage at his lightheartedness. "Can you never take love seriously? For you, it always amounts to sex and seduction, possession and selfishness. You really can't see how limited these versions of love are?"

"I merely refuse to be as hypocritical as most people are in dealing with love," said Thomas matter-of-factly, taking a sip of his water. "The majority of the world pretends that what they feel is a selfless, romantic, spiritual passion that's absent of ego or the desire to possess. I happen to be more honest and self-knowing than most; in standing by my convictions, I take a risk that the world will look down on me—but that's a risk I'm willing to take."

"But haven't you found that that approach grows old—both as an idea and as a practice?" she challenged. "Don't you seek something higher, better, more lasting?"

He replied simply, "These are the limitations we're working with in this world, in this life. We're dealing with human nature, society, and physical constraints. It's useless to protest against them; they merely are."

She pushed away her plate. "I find it inconsistent that where the long seduction is concerned, you believe imagination can transform reality; but where love is concerned, you restrict the imagination."

"I don't restrict it in love either. I merely make a clear distinction between the reality the imagination asserts and the truth of things."

"Well, just as you once said that jealousy is a symptom people read as a sign that their relationship is a success, I believe that romance functions in the same way." She set her chin with conviction. "It may be just a symbol or gesture, but it's no less important for all that."

He considered this. "If romance didn't smack so much of inanity— or if it didn't rely on the attributes people wrongly ascribe to love that I mentioned a moment ago, I wouldn't mind subscribing to it. But until it resembles something closer to the truth, I demur."

Nan remained distant for the rest of the meal, and Thomas avoided saying anything more on the subject. He sensed he had managed to infuriate her, and while this had given him a thrill at the time, upon cooler reflection, he wasn't sure how wise he'd been. Her doubting of his beliefs had shaken his own conviction a little, even as he had outwardly refused to yield. He knew all she really sought was for him to mitigate his extreme position. He didn't quite know what had come over him that evening that had made him so forcefully espouse scientific truth over poetic truth, which was what romance boiled down to in the end. Could it be that the undeniably romantic setting of the lodge itself was making him put up stronger defenses against romance?

As they sat on the couch finishing their wine in front of the dining room fire after dinner, he tentatively wrapped his arm around her and was relieved when, after a moment's hesitation, she responded by nestling snugly into his chest.

"Where are we headed to tomorrow?" he asked.

"I thought we could ski some of the lodge's trails." She showed him the map she'd picked up from the front desk. "If we take this outer route up to Big Baldy Peak, we should get a good portion of our climbing over with in the first hour or so. Then, after we've come down Otter Slide, we'll get a chance to warm up again, with a gradual climb back up to the lodge. Mary, at the front desk, was saying that the actual moving time for a fit person to do all this in would be a little over four hours."

A little later, as they removed their clothes for bed, Thomas felt unaccustomed discomfiture as he placed a hesitant hand on Nan's waist and caressed her gently. The thought had steadily crept through his mind that it hadn't been worth it to maintain his smugly rakish posture at dinner, if he had hurt her or them in the process. In hindsight, he suspected he didn't have enough courage to persevere in cold-heartedness or full-on skepticism—at least not in the face of such conflicting feelings as those that now confronted him. Tenderness, remorse, humility, and even heartache washed over him in rapid succession, and he instinctively bowed his head in recognition of their power. He didn't know what to do about these feelings, other than show Nan through his gestures how contrite he was. He was overcome with the kind of shame he had felt when, at age five, he had deliberately toppled his older brother Andrew's laboriously constructed card house—which had towered impressively above Thomas's own head. The split-second decision to destroy the house yielded the briefest moment of elation, which dissolved instantly into guilt and a sense of utter aloneness. That delight in destruction then, like this posture of heartlessness now, gave the most fleeting of highs before delivering the most abysmal of lows.

Now he tried to convey to Nan, through his eyes and his bearing, how docile and penitent he was. It wasn't difficult for him to find the actions to express his feelings, sincere as they were, but he wasn't sure how she would receive them. He touched his fingers softly to her hair and took a step closer to her, looking meekly into her eyes. He read in them forgiveness, understanding, and love, and for these he was moved beyond words.

♩ ♩ ♩ ♩

As they put on their skis the next morning, they both joked about replacing the soreness they felt from the snowshoeing of the previous day with a new kind of soreness they'd acquire from today's skiing. Nan, who'd grown up cross-country skiing in Minneapolis, showed Thomas the basic gliding motion and the correct way to do double poling and diagonal poling. She also showed him the main methods for getting up a steep track—shuffling, herring-boning, and side-stepping. Then she demonstrated the half wedge and double wedge for slowing down one's descent of a steep hill.

"But keep in mind," she warned, "skiing on the flat prairie, it was rare that I encountered the kinds of hills that you and I may tackle today. So I'm going to be careful, especially on the downhills."

As they climbed the steep section to Big Baldy Peak, Thomas fell backward into the snow and burst out laughing as he imagined how helpless he looked. Nan joined in his laughter before talking him through how to get his skis parallel in the air again, and then how to roll over and stand up.

"That's my ab workout for the week." Thomas dusted the snow off his jacket.

Luckily, they had only a short space to cover before they emerged at the top of the ridge, from which the views were spectacular: to the West, they looked down on Redwood Mountain, and to the North, they could make out Buena Vista Peak, which was nearly as tall as Big Baldy. As he kissed her, Thomas felt Nan shiver.

"Let's get moving again," he said. "At least we should have shelter from the wind once we descend towards the East."

As when they rode their bikes, Thomas took the descent much more confidently than did Nan, but soon they reached some flatter portions of track that they could both enjoy. The route became most fun as they handled the turns of Tower Run, where Thomas began to feel comfortable and even deft at shifting his weight from leg to leg. It started to snow as they approached their climb up to Chimney Rock, and the intense wintry whiteness, in Thomas's view, heightened their intrepidity. They paused at the top, and Thomas noticed that Nan looked nervous as they peered down Otter Slide.

"I think I could relax a little more if there weren't such a sharp turn at the bottom," she said doubtfully. "I always prefer descents where I can see a straight flat portion up ahead."

"Do you want to take off your skis and walk down?" He wrapped a portion of her scarf back around her neck where it had slipped loose.

She shook her head. "It's almost a mile long. I want to try skiing it."

Thomas decided to remain behind Nan rather than taking the run fast as his instinct prompted him to do. She apologized several times for her slowness and caution, but he was glad she seemed to have found a safe and effective method for negotiating the slope. After the first third of a mile, the track suddenly became as steep as some of the most challenging descents on their bike rides. Nan was unprepared for the change in grade, and she let out a cry of alarm as she began to slide out of control down the trail. Then her cry became a scream. As if he saw it unfolding in slow motion, Thomas saw Nan panicking and heading off the track into the snow bank towards a gigantic tree to their right up ahead. As quickly as he saw this happening, he made the split-second decision to take the trail faster than she was travelling off of it and bend around at the last moment crosswise into her path to block her trajectory. It proved to be a little trickier than he thought to angle himself off the track into the snow bank, but he made it just in time to check Nan's skis perpendicularly a mere foot from the tree's trunk. Their forceful collision knocked both of them off their feet sideways into the snow, and Thomas became instantly aware of a painful stinging in the side of his arm where Nan's pole had jabbed it during the crash. He also began to feel a dull pulsating in his upper thigh where he'd landed on a submerged tree stump. But after a few seconds of heavy breathing in the snow, Thomas was relieved to look around and see Nan stirring about ten feet away.

"Are you okay?" he asked.

"Mmm. I have no idea where my other pole went," she mumbled, groping around in the snow.

"Let's focus on getting up and making sure we're all in one piece. Then we can worry about the pole."

They painstakingly went through the process of standing up again, and they found Nan's pole behind Thomas. After they'd dusted each other off, he said, "Did you strain or tear anything?"

"I don't think so. I just feel a little dazed, that's all. What about you?"

He couldn't detect anything worse than the odd sensations in his thigh and arm. "I'll make it back fine. Do you want to walk the rest of the way down?"

"I think we'd better," she said with regret. "I didn't want to have to say I'd walked any part of this route today, but I guess I'm humbled now."

They had only another third of a mile to go before they reached the sharp curve at the bottom of the hill, and then they skied the slow, steady climb back to the lodge without further incident. After they'd returned their skis, poles, and shoes to the rental hut, Nan sighed with pleasure.

"There is no more wonderful feeling than removing ski shoes at the end of a long run."

Thomas had begun to feel more acutely the pain in his thigh and arm. The sudden increase in pain reminded him of those times, racing the bike, when adrenalin kept him from feeling the effects of minor crashes with other cyclists until the end of the race. Then, invariably, the extent of any physical damage slowly dawned on him and he marvelled at his body's ability to push through that pain to fight for a victory—or, in some cases, for a mere finish.

As they removed their gear in their cabin, Nan gasped when she saw his leg and arm. The abundant blood from the cut on his arm had long ago congealed, and the effect was grisly. Meanwhile, two large purple bruises graced his upper thigh on the other side.

"How did you ski through all that?" She grabbed a cloth, applying some peroxide to it, and gently patted at the wound in his arm.

He winced from her ministrations. "I hate to say it, but this is nothing compared to some of the crashes I've been through while racing."

She had a soft, adoring glimmer in her eyes as she cleaned his wound. "But this time you didn't have to crash. You could have let me hit the tree. I still don't know how you managed to block it."

"I don't quite know how either." He relived the scene in his mind's eye. "I think we were lucky the snow off the trail was so deep that you couldn't travel as fast as I could on the trail."

She kissed his nose, and he pulled her down to meet his mouth in a deep kiss. "This I will say. In all my racing career, I've never had such an entrancing nurse for my wounds."

She laughed. "I'll bet you say that to all the women who take care of you."

He kissed her hand, his eyes sparkling. "I love *this* woman. That lends special weight to my words."

After she had placed bandages on his arm, she made up a small bag of ice for him to hold against his thigh. She tucked him into the bed and said, "I'll mix up the whiskey sours tonight. I've seen you do it—I know how."

From the warmth of the blankets, Thomas watched as she vigorously shook their drinks and strained them into the glasses, adding a cherry to each one. She climbed in next to him and they raised their glasses.

"Thank you for saving my life," she said. "Again."

"Any time." He smiled. "I'm already gearing up for our next disaster, whatever form it may take."

Thomas was lost in thought as they drank their cocktails. Was he wrong, after all, about love's essential self-interestedness? Had he only blocked Nan from the tree so that he wouldn't have to feel guilt later over having failed to do so? So far as he could remember, he had barely even thought before acting. Surely, in order to have been motivated by self-interest, he ought to have calculated the pros and cons of doing what he'd done before he did it. He was struck by the spontaneity and forcefulness of his impulse to save Nan—was that, after all, what it was to love someone? If so, he had falsely maligned love. It had been one thing to pull her back from her manic dash after the thieves a few weeks before. That had been easy, painless, and practical. It felt like quite a different thing to lose himself in the face of danger this afternoon. Until this moment, he would have said that in attempting to save her he was merely saving his extended self. Now, he realized he had completely put himself aside to rescue her. He had unhesitatingly forgotten himself—much as he remembered his dad doing one time when he was six years old and learning how to ride a bike. He was about to fall on the concrete, and his dad had thrown himself between him and the ground, breaking his fall. The resemblances between the two scenes were startling, and he discovered with a jolt that his love for Nan was much more like his dad's own love for him than he would ever have admitted before today.

He flushed with embarrassment to think of the assurance with which, only the evening before, he'd upheld his stance regarding erotic love. It almost seemed as if today's adventure had been deliberately laid out to mock and humble him. He knew Nan was far too magnanimous ever to mention the irony of the situation, but he felt that in order for the experience to be complete, he must.

He swallowed before speaking. "I, um . . . I think I owe you an apology."

She looked surprised. "Whatever for?"

"For doubting your wisdom last night," he said, still proceeding with uncertainty. "I believe I may have misjudged the power of love after all."

She remained silent, waiting for him to continue.

"I guess today tested my long-held theory and it failed to hold up to experience. I now begin to believe that erotic love can be selfless after all—or perhaps, that part of it that resembles what two close friends share, or what a parent feels for his child." He paused a moment, and then determined to speak aloud his other thought. "I begin to wonder what else I'm wrong about where love is concerned."

She chose to enlarge upon his first reflection. "This was the first time I personally could vouch for the truth of what I've read in books and believed merely in theory for so long, so I can hardly blame you for holding out so long for definitive proof."

"So you find it satisfying to see it play out in real life?"

She nodded. "Part of me agreed with what you said last night about the troubadours and medieval romances. I've always been secretly frustrated that more people don't acknowledge the self-interestedness of such love. All the same, I'm too romantic to hold out long against disinterested love. I know that the love my mom and dad have always showed me occurs all the time between married couples—starting with my parents' own relationship."

He smiled. "Maybe we should have stuck to Shakespeare after all, in our conversation last night. I seem to remember Desdemona blaming herself for her own murder and telling Emilia to recommend her to her kind lord, Othello. That's pretty selfless."

"And then there's Viola," Nan added, "who declares she would gladly die a thousand deaths for Count Orsino."

"Perhaps Shakespeare didn't believe so much in selfless men as in selfless women," Thomas suggested. "I begin to see why he leans in that direction. Women are far stronger than men. And it takes great strength to be selfless with any constancy."

Nan replied generously, "If women have the strength and constancy, men have the courage and drive. It seems like a perfect pairing."

Thomas placed his hand at the base of her neck and leaned in to kiss her. "I can't disagree with that sentiment."

After they'd kissed, Nan smiled slyly.

"What is it?" he asked.

"I just realized you can't have meant everything you said against romance last night," she said. "For you once said you admired the French notion that love consists in sharing the small routines of everyday life. What can be more romantic than that?"

♪ ♪ ♪ ♪

It continued to snow heavily all night and into the next morning, and at breakfast on Sunday, they heard from other diners that chains were now being required on all vehicles.

Over their plates of sausages, bacon, eggs, and English muffins, Thomas and Nan discussed how best to use their last full day.

"I imagine you want to avoid putting chains on your car at all cost," she said. "If you're up for it today, we could ski across the road on that Upper Woodward trail that Julie mentioned. Maybe by tomorrow they'll have cleared the road enough that you won't need chains."

He had been thinking along the same lines. "How's your body feeling after the skiing of yesterday? Surprisingly, mine isn't as sore as it was after the snowshoeing."

"Yeah, I don't understand that either. Maybe the upper-body action of the poling takes a lot of stress off the lower body."

Thomas could tell that Nan felt more comfortable on today's trail than she had felt on the trails of the previous day. The ascents and descents were much more gradual, and they had constant astonishing views of sequoia groves rising above them on both sides of the track. After four hours of skiing, they had only just reached the fork where the trail split off into a loop around a mountain that was almost as high as Big Baldy. It had stopped snowing about an hour before, and the fork seemed like an ideal spot to bring out their thermoses of coffee and the peanut butter and jelly sandwiches they'd made up at breakfast.

Thomas took a swig of coffee to wash down a bite of sandwich. "I find it interesting that although neither you nor I is a team player or tends to gravitate towards teamwork, we both seek out occasions where we can join like-minded people to work towards a common purpose. I imagine that's one of the things you enjoy when you play with other musicians or go on your hikes. I certainly love that aspect of riding the bike with you or conversing with my language groups. And here we are struggling together through an intimidating landscape that nearly licked us yesterday. Do you think any of what you and I do resembles teamwork?"

Nan smiled, picking up on Thomas's first statement. "I don't mind team*work*. What I shy away from is team play. In teamwork, everyone still gets to act independently, but when the going gets tough, the team gets things done collectively that individuals could never accomplish. Teamwork involves tackling a great project or task. Team play is different. I rarely feel that games, or even team sports, have an aim large enough to make them worth playing, and I think my skepticism shows immediately when I join a team."

"But is there no way in which our partnership operates like a team?" he persisted.

"I think it's a matter of perspective. If we see something as a worthy task, then we're working together as a team. If we see it as drudgery, then we're merely two fellow sufferers who happen to share one another's company."

He laughed. "An apt metaphor for life . . . and other things."

"Speaking of other things," she said. "I have an idea for tonight. It's actually a game."

"Oh?" Thomas was equal parts surprised, delighted, and intrigued. While he never ran out of ways to vary up their seduction, he loved the thought that Nan had taken the initiative and come up with something creative on her own. "Tell me about it."

She shook her head with an impish smile. "I'll wait until after dinner."

They returned famished from their eight-hour ski run, and wolfed down everything that the lodge's dining room served. Thomas bought, at a rather steep price, some of the lodge's white wine, and once more they sat on a couch by the fire drinking the last remnants of the bottle. They had come back at the cusp of twilight, and Nan had sighed at the beauty of the canyon yawning beneath a waning gibbous moon and a billion brightly twinkling stars that seemed at once frighteningly and intimately close.

On the sofa, Thomas gathered Nan securely in his arm. "Now will you share your idea for our game?"

Nan put her finger over her lips and smiled, nodding over towards the adjacent sofa, where three teen-aged girls had settled and were looking at something on one of their phones and tittering with glee.

Thomas chuckled. "Maybe speaking your idea aloud will dispatch them sooner rather than later."

"You are cruel!" she objected. "It's everybody's dining room."

"It's you who are too kind," he corrected. "At any rate, I'm betting your idea is less risqué than what they're looking at."

"Don't bet too much on that."

"Now you really have me in suspense."

Eventually the three girls were coaxed over to the foosball table by a tall gangly boy and another boy who looked like his younger brother. Thomas wondered, not for the first time, if he had looked as ugly and awkward at the same age. No doubt he had.

Once they were alone again, Nan said in a low voice, "It's a truth game I devised while we were skiing out to the fork today—it's kind of a variation on truth or dare and two truths and a lie. The question-asker poses a question to the question-answerer. The answerer can choose either to lie or tell the truth. Then the asker must decide which one the answerer has opted for. Each time the asker gets it right, that's a point for the asker. Each time the asker gets it wrong, that's a point for the answerer. At the end, the person with the fewest number of points has to devise a scenario that's calculated to give the most pleasure to his or her opponent."

Thomas said, "What if what gives the most pleasure to your opponent is also what gives the most pleasure to you?"

"That's all right." Her eyes glittered. "At that point we're working as a team to make the game as universally fun as possible."

"This must constitute a worthy task in your books." He smiled. "Do the players alternate regularly with their questions, regardless of whether they earn a point or not?"

"Yes. That should keep it interesting."

"Oh, it sounds interesting enough." He ran his free hand down her sternum and over her belly to her groin. He was, as always, gratified to feel her shudder beneath his touch. "It has all the elements we love most: competition, risk, and creativity."

Then she said wistfully, "What a shame we can't bundle up under those gorgeous stars and play the game. Even under all the down blankets, I don't think we'd last more than a few minutes."

"We could try to play it in the hot tub," he suggested.

"What a brilliant idea!" She kissed his cheek euphorically. "Let's go."

The stillness of the night and the majesty of the multitude of stars that spread in a canopy directly above them silenced them both for a while as they acclimated their bodies to the hot water.

Then he said, "Since you came up with the game, why don't you ask the first question?"

She paused a moment before asking, "Why did you wait so long to do the long seduction, and why did you choose me?"

He laughed. "Isn't that two questions?"

"You've given me reason to believe they're closely related, so you might as well handle them both together."

"I think my tendency to cheat is wearing off on you," he said merrily. Then, more soberly, "Well, here's my answer. I've always tended to take charge in relationships, with greater or lesser success. Some time in my early twenties, I began to realize that I succeeded best with women who appreciated danger. But that quickly became too predictable for me. I saw that to keep the seduction interesting, some further intellectual element was necessary. I decided that, ideally, both partners doubt the truth of the roles they're playing and doubt their partner's commitment to the truth. I found it was easy to find jealous partners, but it was extremely hard to find partners who could hold two contrasting ideas in their mind at the same time and grapple with competing truths. Out of boredom with the sexual relationships I had, I became more and more dedicated to the idea of restraint. Had you proposed to my twenty-five-year-old self the idea of delayed gratification, he would've scoffed you out of the room. But at thirty-five I made that a condition for dating any woman—how well and how long she could contain her desires. Unfortunately, as a result, the relationships I had over the last few years were like three-legged tables with only two legs. If a woman had sexual discipline and a sharp mind, she didn't care for danger. If she had willpower and a taste for danger, she didn't have the capacity for doubt. So I abandoned the idea of looking for all three, and settled for just two."

"And then?"

He smiled. "I think you know how the rest goes. From day one, our conversations and your demeanor told me a great deal about you. I broke up with Marina six months into knowing you because I found I was getting more out of our rides than I was from my relationship with her. I began to suspect that, after all, I could have all the three elements I sought in a relationship. That's when I fleshed out my ideal version of love, and the concept for the art of the long seduction was born."

"But how much longer were you willing to wait before you made a move with me?"

"I'll ignore the fact that you've now asked three questions." He chuckled. "I was going to give it until the year mark—for if you'll recall, we met on December 2, 2020—and then I was going to ask you out."

Her jaw dropped. "You remember the date we met? You were only going to wait another three weeks? That's so romantic!"

He flushed. "That's why I avoided mentioning it."

"And all three elements of the seduction were present with us?"

"Well, the hardest part for me—and perhaps for you too—was the very same restraint I had made a basic condition of previous relationships. That still is the most challenging part."

She seemed to recall that they were playing a game. "I'm going to guess that everything you've just told me is the truth."

He laughed. "You can't really make this kind of thing up."

"Your turn," she said.

"Why do you avoid all formal competition, when you clearly love competing?"

She took a deep breath and looked down at Thomas's chest as she spoke. "When I was young, I often frightened myself with the levels of competitiveness I showed. I alienated several friends, pushed myself to ridiculous levels, and readily accepted the mantle of head-nerd at school in the course of striving to outperform everyone else in academics, extracurriculars, and athletics. Of course, as I got closer to college, the competition became stiffer and the stakes higher. It was only once I reached graduate school that I decided at last to compete solely with myself. I had to make a forceful mental leap of throwing off the urge to compete with others. For those same reasons I forbade myself from bike racing after tackling it just three times. I came to love a saying New York cyclists have: 'Ride your ride.' It expresses the idea that as a rider you will always be passed up by other people who have different goals and who are at different points in their ride. When you're passed up, instead of getting sidetracked and trying to speed up, you need to keep your sights on your own ride. It seemed like a fitting motto for life."

He nodded. "That explains a lot. You're so fiercely competitive that in order to have any peace of mind, you have to swear off competition. But since you got in three questions, I'm going to tag on another one here: how is it you can enjoy the competition you and I so frequently engage in?"

"I crave the friendly competition we keep up on our rides because it brings only benefits, and we're so well-matched. The same thing applies to our conversational banter. As far as the foreplay and sex goes . . . I'll confess I have often wondered what happens if I push you to the limit physically. Sexually—and, perhaps, emotionally—you have me wrapped around your little finger. But what if I really fought with you? What then? The idea greatly arouses me."

Thomas felt his own keen arousal at the image and he brushed back a few strands of Nan's wet hair. "I've been waiting to hear you say that. It definitely increases the danger factor of our lovemaking. But there are ways we can keep things basically safe—just as wrestlers have safe techniques for practicing with one another. Yet it also tests our trust of each other."

"Perhaps we can start by arm wrestling," she suggested. "You know how I love your forearms."

"We seem to be straying from our present game. I'm going to guess that you have also just given a truthful answer to my question."

"We may both be too honest for this game. We've each got one point, I think." She ran her fingers over his chest and down to his erection, which she massaged gently in her hand as she asked the next question. "You once asked me what I'm most afraid of losing, and I said my freedom. What are *you* most afraid of losing?"

"With your hand where it is right now, I can't guarantee I can give you an accurate response." He emitted an involuntary moan. "Before I forget, I'm going to answer now—my reason."

She smiled. "That's a nice irony. But do I really believe you? I'm inclined to think you do fear losing your reason most of all. In losing your reason, you lose your memory, your control, your discipline, your will, your analytical abilities—all things I know you cherish. I'm guessing you've just spoken truthfully."

"I have," he said shakily as she moved her hand down to his balls and played with them.

He groaned again, and she continued to muse aloud.

"I'm wondering what the ultimate danger would be for you in our seduction that equates with the danger I always confront? For you know I fear losing my freedom—yet I continually cede it to you when we make love." She perched on his thigh and lingeringly ran one hand up his length, as she cradled his balls with her other hand. "What if I should tease you beyond your capacity to think, and then hold you to a very important decision in the midst of my teasing? I don't think that would be so very difficult, given what I know about your preferences. For I too have been observing you. And timing has always been one of my strengths."

He closed his eyes and tried to concentrate on something far-removed from this seductive tableau, but all he could think about was Nan's graceful fingers handling his manhood. Then he imagined those same fingers dancing deftly over the keyboard as she practiced her pieces. He saw them scrawling notes in her staffed notepad, and carelessly scouring his frying pan, and closing her ears against the loud sirens in the street near his bar one night, and ingenuously reaching to cover his eyes on the bench, and skillfully guiding the pen as she sketched his portrait, and he was flooded with overwhelming desire for her. He no longer wished to fight back against the tide of love that propelled him towards her. He doubted he was even capable of doing so. He found himself blithely abandoning all pride, poise, and presence of mind as she wove her spell over him.

"For now, I think I can persuade you to yield complete power to me tonight," she said in her low voice, laying a slow trail of sensual kisses from his shoulder to his neck, but not letting up on her sweetly torturous teasing of his genitals. "Are you prepared to do that?"

His mind was in a haze and when he opened his eyes, he saw her at first indistinctly. But her voice had clearly asked him a question, and he registered it as one he could answer simply.

"Yes."

"You're sure?" She straddled him now and wrapped her legs around his waist as she interlaced her fingers behind his neck, her warm softness enveloping him like feather down.

His eyes fastened on the bewitching décolletage that her bikini showed to marvelous effect, and then they wandered up to her lovely neck, her expressive lips, and her brilliant eyes. Who would be anything less than a willing captive to these exquisite charms? He fancied he had become the supple clay beneath her expert sculptor's hands, and the image fed his tractability.

He nodded, now more than ever forcibly convinced of his humbling. How could he refuse her anything she asked for? He was enraptured by her. He would gladly lay down his life for her. He was utterly hers. If it made him into a blithering fool, then so be it. It was well worth being a fool for such enchantment. Reason be damned.

She seemed to read in his eyes the adoring thraldom she sought, and she brought her lips closer, hovering just an inch away for what felt like an eternity.

"I'm going to hold you to your promise."

Then, at last, she kissed him.

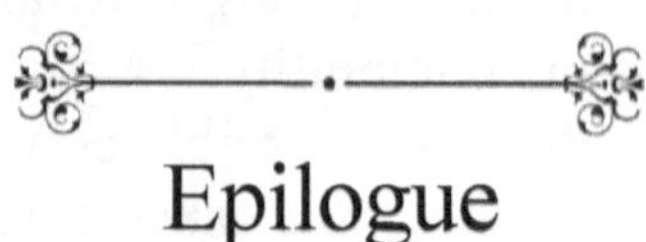

Epilogue

Dennis handed Thomas the pen for signing the guestbook and then took a drop of hand sanitizer and rubbed it into his palms. "This is a really elegant venue," he remarked.

"I didn't even know this existed until I Googled it after Teddy told me about it," said Bess effusively. "It's so light and airy."

Bertie had rolled up one of the paper programs in his hand. "Uncle Teddy said I'm at a kids' table."

"That's right—this card says you're at Table 4," said Bess.

After Nan had written a message for Teddy and Naomi in the guestbook, she and Thomas followed Dennis, Bess, and Bertie into the reception room, where a few dozen people moved about with drinks or stood in clusters talking.

"I'm headed to the bar—what can I get you all?" Thomas asked Dennis, Bess, and Nan.

"I'll come with you," Dennis offered.

"White wine, please," said Bess and Nan at the same time, and then they smiled at each other.

While the two men were gone, Bess and Nan stood in the center of the room looking about for relatives or friends they recognized.

"How did the interview for the adjunct position at Diablo Valley College go this week?" asked Bess.

"Really well," said Nan. "They hired me to teach two courses—one in music theory and the other in composition. Classes start up in late August."

"It's your website." Bess nodded. "It really showcases your accomplishments. I'm glad you decided to have some musicians perform your quintet early so people can listen to it now."

"I got a slew of new subscribers to the YouTube channel after I put that up—and lots of positive feedback."

"Will you still give piano lessons?"

"Yes, I'm keeping all seventeen students I currently have. I'm giving most of my lessons on the weekends."

"So no more dogwalking or substitute teaching?" Bess smiled.

Nan shook her head, matching her smile. "Not for now."

Lewis and Lilian came over to them and exchanged full, warm hugs with them.

"I loved the music you played for the ceremony," said Lilian to Nan. "What a creative idea to play Debussy's Prelude number one as Naomi walked down the aisle!"

"I liked the composition you wrote for the bride and groom," said Lewis. "It had a bit of a New Orleans sound to it in the middle."

Nan beamed. "I love Jelly Roll Morton. It must've been his influence you were hearing."

Thomas and Dennis returned with the drinks, and Frank joined their growing circle.

Nan sipped her wine. "Naomi and Teddy couldn't have ordered up a more perfect spring day for their wedding."

"Believe it or not, this Minnesota sun is a reprieve for us Californians," Dennis commented. "It's cold and pouring rain in the Bay Area right now."

"Better that than the extended heat wave they're having in the Pacific Northwest." Frank shook his head. "Like your wildfire season out West, summer seems to have been moved up several months now in all too many places."

"And don't forget the Midwestern tornadoes of December," said Bess somberly.

"The time is out of joint indeed," Frank agreed. "But at least now that there's no longer any room for doubt as to the role people play in climate change, that frees up policymakers like yourself"—he turned at Dennis—"to make the economy cleaner."

Dennis nodded. "Unfortunately, the knowledge is a double-edged sword. It makes some people want to give up before we've even tried."

Thomas said, "I guess we also have to factor in inconsistency. If individual humans are inconsistent, they can hardly expect their governments to be consistent."

"True." Dennis smiled wryly. "That's another of our biggest challenges."

Frank turned his glass around thoughtfully. "They've almost completed that steel vault in Tasmania that holds all the information for future generations in case this one gets wiped out. But I think the vault is more symbolic than anything else—a way of warning people of the worst-case scenario."

Lewis observed, "It's not human nature to keep in mind the worst-case scenario for very long—not for oneself, at any rate. Here we are, at Teddy and Naomi's wedding, and we, of course, want to believe that their children can live on earth for a little while."

"And their children's children," added Lilian.

"Speaking of the bride and groom . . ." said Frank, as a beaming Teddy and a radiant Naomi entered the room and began to circulate among the guests.

Thomas placed his hand on the small of Nan's back and leaned in a bit closer to her, saying in a low voice, "You looked beautiful at the piano. And you sounded good too."

"You don't think anyone could detect that the piece I composed for Teddy was unfinished?"

Thomas shook his head. "To me it seemed complete enough—if that's not an oxymoron."

Nan tilted her head. "That idea could be the basis of a poem."

"You haven't written as many of those recently."

"No. Music has kind of taken over for the time being. But I love it."

After a rocky start to the year, Nan had spent the early spring throwing herself into preparing for her Perseid group's March performance of the Schumann, Brahms, Debussy, and Grieg pieces, which they had recorded and put up on their YouTube channel. In February she'd gotten a San Francisco-based chamber group to perform her own quintet, and had consequently posted that recording on the channel. Around the same time she'd also convinced Albert to record the Franck sonata, which she'd included on her website. Meanwhile, she'd composed a good chunk of a song cycle for mezzo soprano and piano.

In the midst of all this musical productivity, she'd begun to worry less about spending enough time on her other artistic and literary pursuits. Paradoxically, now that she just engaged in these passions whenever the spirit moved her, she found the process less mechanical and more organic, even as the fruits of her efforts were, to her, more satisfying. It had taken the leap of faith in November and December for her to overcome her former uncompromising attitude towards the pursuit of excellence. Being in love and being committed to her composition had started Nan on a path towards living more fully in the present, so that she could enjoy and make the most of time. What began as a daunting puzzle for her had now become a comic challenge: how to harness time in both her own art and the long seduction so that doubting became a creative impulse. In this way, she had cured her more general obsession by means of two particular obsessions— her composing and Thomas. Luckily, both obsessions promised to yield infinite variations in times to come.

Something Thomas had said to her on their drive back from Kings Canyon had particularly helped her relax into her new habits and behaviors:

"You may not think you're changing from day to day, but change will *happen* to you—like it or not—and one day you'll no longer recognize yourself."

Seen from that angle, loss of control seemed an inevitable human process that led to growth and change.

Having given herself over to love, Nan cherished Thomas's controlling nature, his cruel and selfish streaks, his intransigence, and even his indifference to music. She adored the whole of him, because he was Thomas.

Now he said, "You know I consider myself a reluctant romantic at best. But this wedding so far is testing my skepticism."

"How so?" she asked.

"The simplicity of the gestures was most striking. The vows were completely unpretentious—your brother and Naomi merely expressed how much they love each other and are committed to one another—and the ceremony didn't have anything extra or over-the-top. Even the style of dress and décor were minimalist."

"I do recall you once advocating simplicity in marriage proposals," said Nan in a sprightly tone. "Does this mean you're a new convert to romance?"

"Since loving you, I can no longer afford to disparage it as I once did."

"That's a step towards belief." Nan placed her arm around Thomas's waist. "And where do you now stand on the social trappings of romance?"

"You no doubt refer to cohabitation, marriage, and children. But you'll recall I was never averse to those." Thomas regarded her fondly. "If anything, I believe it was you who once said that convention was the very worst reason for doing something yourself. Doesn't that bar you from these domestic practices?"

Recognizing that she had been routed, Nan smiled self-mockingly. "I confess that individual lived experience is gradually teaching me the wisdom of certain conventions."

Thomas's eyebrows rose quizzically. "That sounds to me like the greater conversion. I must conclude from your example that Benedick was right when he said that man is a giddy thing."

"Like him, I'm not ashamed to own it."

THE END

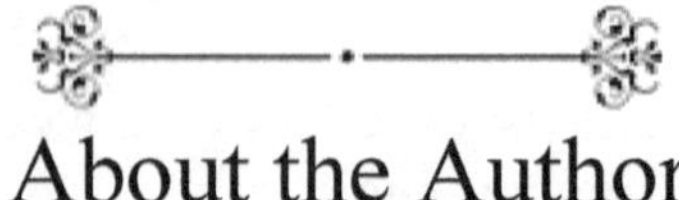

About the Author

Kathleen Haley is a writer, scholar, and educator in Berkeley, California. Her primary area of research interest is humor across different cultures of the world, and her favorite genre of novel to write is contemporary romance. She has written three such novels—*Rivals in Restraint* (self-published), its sequel, *A Perfect Understanding* (currently in submission), and *Under the Olive Groves* (currently in submission). She lives with her dachshund-Chihuahua mix Pooh-bah, and when she isn't writing, she loves to cycle in the nearby hills, lift weights, and play piano.

www.ingramcontent.com/pod-product-compliance
Lightning Source LLC
Chambersburg PA
CBHW030921300726
48970CB00001B/260